ANATOMY OF THE PASSIONS

Cultural Memory

in

the

Present

Mieke Bal and Hent de Vries, Editors

ANATOMY OF THE PASSIONS

François Delaporte

Translated by Susan Emanuel
Edited and foreword by Todd Meyers

STANFORD UNIVERSITY PRESS

STANFORD, CALIFORNIA

Stanford University Press

Stanford, California

© 2008 by the Board of Trustees of the Leland Stanford Junior University.
All rights reserved.

Anatomy of the Passions was originally published in French in 2003 under the title
Anatomie des passions © 2003, Presses Universitaires de France.
Additional material has been added.

Foreword © 2008 by the Board of Trustees of the Leland Stanford Junior University.
All rights reserved.

Printed in the United States of America on acid-free, archival-quality paper

Library of Congress Cataloging-in-Publication Data

Delaporte, François, 1941–
  [Anatomie des passions. English]
  Anatomy of the passions / François Delaporte ; translated by Susan Emanuel ;
edited and foreword by Todd Meyers.
      p. ; cm. — (Cultural memory in the present)
  Includes bibliographical references and index.
    ISBN 978-0-8047-5850-5 (cloth : alk. paper) — ISBN 978-0-8047-5851-2 (pbk. : alk.
paper)
    1. Expression. 2. Facial expression. 3. Emotions—Physiological aspects.
4. Physiognomy. I. Meyers, Todd. II. Title. III. Series.
    [DNLM: 1. Facial Expression. 2. Emotions. 3. Physiognomy. WE 705 D338a
2008a]
BF591.D4513 2008
153.6'9—dc22

                            2007044958

Typeset by Bruce Lundquist in 11/13.5 Adobe Garamond

To Monsieur Bernard Devauchelle, surgeon, in grateful recognition.
—François Delaporte

It is always a matter of unseating the Ideas, of showing that the incorporeal is not high above, but is rather at the surface, that it is not the highest cause but the superficial effect par excellence, and that it is not Essence but event.

—GILLES DELEUZE

Contents

Foreword

SPEECH WITHOUT UTTERANCE

The study of facial expression and its musculature undertaken by Guillaume-Benjamin Duchenne de Boulogne in 1862, something he described as his *orthography of facial expression in movement*, was an attempt to secure biological meaning in the natural language of the emotions.[1] Duchenne believed in the existence of a language of emotions, the *passions*, as they are revealed through the signs of facial expression. The belief was certainly not new: the study of the passions was an elusive subject stretching back to René Descartes's philosophical writings on the dimensions of the soul (1649) and Charles Le Brun's aesthetic teachings on emotion in sculpture and painting (1667).[2] A decade after Duchenne's photographic studies, Charles Darwin would use Duchenne's work as a touchstone for his *The Expression of the Emotions in Man and Animals* (1872).[3] However, Duchenne did not simply move the disciplines of myology and physiognomy forward; he opened and reshaped their dimensions.

For the philosopher Emmanuel Levinas, the face is a thing that "resists possession"; expression, while "still graspable, turns into total resistance to the grasp," its meaning made available "only by the opening of a new dimension."[4] That meaning in expression requires such an opening echoes Duchenne's notion of the universal or at least taxonomic attributes of facial expression. Duchenne, through techniques of localized electric stimulation and photography (and, perhaps more importantly, attention to the *movement of signs* in the living body), breaks through the form of emotion—the surface and the *underneath* of the face—that nevertheless delimits it.

Through Duchenne, François Delaporte provides a remarkable philosophical and historical examination of expressive physiology during the mid-nineteenth century, considering the science of emotion as a means of revealing inner life—thoughts, feelings—upon the surface of the face.

The central concern of Delaporte's study is how techniques of study-ing facial musculature became a point of contact between existing and novel understandings of the body's expressive anatomy. Delaporte shows that Duchenne did much more than contribute to customary knowl-edge; instead, he entirely reordered the knowledge and limits of expres-sive physiology in science and art. The face became a site where the signs of inner life are silently revealed, not yet betrayed by speech but brought forth by reflexive physiology or by technical manipulation.[5]

Delaporte's essay on Duchenne shows the close connection between attempts to understand the "vocabulary of nature"—complete with the difficulty of meaning, absorption, and translation—and the vocabulary of experimentation.[6] Duchenne uses a quote from Francis Bacon to pref-ace his work: "Experimentation is a type of question applied to nature to make it speak."[7] Later in the same passage of *Mécanisme de la physionomie humaine*, Duchenne refers to Georges-Louis Leclerc Comte de Buffon when he writes that the living face is the site "where each movement of the spirit is expressed by a feature, each action by a characteristic, the swift, sharp impression of which anticipates the will and discloses our most secret feelings."[8] Through the use of electrical currents, Duchenne claims to have "made the facial muscles contract to *speak* the language of the emotions and the sentiments" (emphasis in original). This relationship between the lan-guage of nature and the language of experimentation, as Delaporte dem-onstrates, allowed Duchenne to trace out a physiological grammar of the emotions captured through photography.

The face is the locus of Delaporte's essay, but to say that his study is limited to the face is somewhat misleading. It is equally misleading to say that Duchenne's work is another example of science progressing in the shadow of a chance alignment of ideas and technologies.[9] Delaporte draws the features that bring the face to the center of a constellation of concepts and techniques. Moreover, he shows that Duchenne did not sim-ply combine the old with the new; rather he created an entirely different foundation for facial anatomy in the living body. Such close attention to epistemological turns within the history of science has always been the case with Delaporte's work: the recognition of signs and the discovery of Cha-gas's disease; the constitutive relationship between the *Aedes aegypti* mos-quito, yellow fever, and the birth of tropical medicine; and the historical claim to a concept of disease in Paris during the 1832 cholera epidemic.[10] Here, the *opening of a new dimension*, to take up Levinas's phrasing, is a space where Delaporte locates the organization of concepts, or more accurately the reorganization of knowledge driven by technologies such as electrotherapy and photography, not as a way of explaining how Duchenne's work was made possible but as an "epistemological reorganization (*reorganisation épistémologique*) under concrete historical circumstances."[11] Delaporte follows the availability and circulation of concepts as though they were lines on a face.

TODD MEYERS

Acknowledgment

I would like to thank the Université de Picardie Jules Verne for its generous support of this project.

F.D.

ANATOMY OF THE PASSIONS

Introduction

The problem of laughing, Stendhal used to say, "should be written about in anatomy's style and not in the academy's style." The anatomists did not doubt that the laugh, like any other emotion, is dependent upon the body. But they were not able to describe the relations between muscular actions and the expressive surface. Jean Cruveilhier, the first holder of the chair of anatomical pathology, directly testifies to this: "I willingly agree with Santorini that it is almost shameful for the anatomists to find that the muscular apparatus that is entrusted with the expression of the mobile tableau of our passions should be the least well known in all physical economy."[1] Due to its archaic methods, the framework of classical anatomy excluded any discourse on the structure and use of muscles. In order to see the muscles, one had to make incisions in the face of a cadaver: not only did the dissection sacrifice its object of study, it also prevented any analysis of muscular action. The scalpel forcibly destroyed the moving attachments of the muscles to the inside of the skin. Observing these inert, sagging, and mutilated parts did not permit apprehension of either their structure or their function. The goal of anatomists, said François-Xavier Bichat, "is attained for them when the opaque envelopes that cover our parts are functioning only as a transparent veil that allows the whole and the relations between parts to be discovered."[2] To attain this objective, myology had to depend not on dissection of the cadaver but on a technique that would respect the corporeal envelope—vivisection without mutilation. Just when Cruveilhier was deploring the ignorance among anatomists, Duchenne was reaching this goal. By the expression "living myology," he designated the application of localized electrization (*l'électrisation localisée*) to the study of muscles. With his mastery of electrical current, he was inventing an anatomy on the living. His electric scalpel was a sort of Moebius strip that allowed him to move under the skin. Following the interior pathway, just on the reverse side, he was reconfiguring the myology of the face. He offered a new perceptive structure for anatomy: there were no muscular layers, as most anatomists thought, but an ensemble (*ensemble*) of independent muscles. The rest followed: if the movement specific to a facial muscle is its function, that function is *expression*. Turned inside out like the finger of a glove, muscular action shows its effect on the surface in a physiognomic movement: "There is no need of further developments to demonstrate that my research on expressions falls within physiology, since physiology signifies the study of life and the analysis of functions, and since the muscles of the face are almost all specifically designed for expression."[3] To follow Stendhal's recommendation, it was necessary to invent a new style of anatomy and many other things besides. In this book, we are looking at the constitution of the very first knowledge about expression.

The emergence of a new style of anatomy relied on a modification of electrical techniques. Before any research, it presupposed the formation of a vast theoretical and practical field. In the first half of the nineteenth century, transformations in medicine, physiology, and physics had contributed to reinvigorating techniques of electrotherapy. In medicine, the invention of electropuncture was linked to the theme of localizing electricity in injured tissues. In physiology, the first electro-physiological laws were procuring a rational basis for the treatment of paralysis. In physics, the discovery of electromagnetism was being translated into the construction of Volta electric and Faraday magnetic apparatuses that were soon being used in hospitals. At first glance, something akin to *localized electrical stimulation* was becoming possible. One might be tempted to say that Duchenne had found, respectively: in medicine, the theme of mastery of electricity; in physiology, a style of research that tied its object of study to the method that would reveal it; and in physics, the instruments of choice. But this temptation should be resisted: a theme, a style, and an instrument can never define a problem. In reality, Duchenne set aside the practical interest attached to new techniques of medical electricity. In effecting a break with the first therapeutic convictions, he was able to create from scratch a method for exploring the electro-physiological properties of muscles. Through the rectification of a technicality that was too close to the immediate interests of medical electricity, Duchenne fabricated a special tool. Specialization, which is the goal of a scientific orientation, carries interests that are more speculative and more complex. With the invention of a theoretical use of electricity, Duchenne had at his disposal a new tool of knowledge that permitted registering, on human beings, electro-muscular contractility and sensitivity.

The subject of this history does not arise from any previously constituted discipline. By its application to the domain of anatomy, localized electrical stimulation was becoming an instrument of exploration. From it flowed a complete reorganization of the facial muscles. This living anatomy included a physiology allowing researchers to apprehend the signification of physiognomy in movement, because physiological mechanisms were on an equal footing with expressions. By elucidating the correlations between expressive muscles and the play of physiognomy, Duchenne renewed the study of indices. This first theory of expression, because it was linked to the philosophy of the natural sign, also defined a theory of the natural language of passions. Between the soul and the movements of physiognomy, Duchenne slipped a linguistic structure of muscular actions. From the moment when the doctor from Boulogne enunciated the rules of the natural grammar of passions, he posed the precepts of the art of painting them correctly. Since Le Brun, encounters between knowledge and art had been celebrated in so many ways that one must discover what that meeting meant precisely for Duchenne. Around the 1870s, with the collaboration of Mathias Duval, his pedagogic project entered into the teaching at the École des Beaux-Arts. For the benefit of students, several dozen photographic enlargements were glued to canvas and mounted on an oval stretcher.

The description of the stages marking the discovery of how expression functions calls for some precise explanations. The first concerns the manner in which it was necessary to order these stages. Each presupposes a preceding one, but this subordination is not reducible to an order of derivation. Each stage constitutes an event that also designates an opening. The study of the muscles of the face and of their actions depended on localized electrization. However, this anatomical and physiological recasting necessitated the integration of an ensemble of bodies of knowledge that were both neuro-physiological and neuro-pathological. Analysis of the mechanism of physiognomy in movement could not appear until there was a remaking of muscular anatomy and physiology. Nevertheless, this examination that correlates muscular actions and expressions is conditioned by the photographs of electro-physiological experiments. The study of the language of the passions rests on the knowledge of the mechanisms of expression. But the perception of the signs of this language in action carries a symptomatology of the passions and a new profile of what can be perceived and enunciated. Finally, the rules of tracing the expressive lines are founded on the grammar of the natural language of the passions. But it is once this grammar is constituted that gaps or faults in the pictorial language may be identified.

The second proviso concerns the manner of inscribing the constitution of a body of knowledge of expression within the history of ruptures. "The amount of 'progress,'" said Nietzsche, "can actually be measured according to how much has had to be sacrificed to it."[4] In this case, this means the whole tradition that Duchenne's predecessors—indeed, his contemporaries—had inherited, which included: an anatomy and a physiology of muscles of the face that were not different from the ones that Jacques Benigne Winslow had given; a perception of the expressions as intentional language that remained in the lineage of Descartes; Johann Caspar Lavater's pathognomy without any organic support; and finally, in academic teaching, the norms of pictorial language illustrated by the schemas of Le Brun. Here arises the third proviso, relating to periodization or, rather, to the various recurrent terms. The names of Winslow, Descartes, Lavater, and Le Brun are valuable clues here: it is as much a matter of relating the history as describing the actuality of the moment when the function of expression comes to light.

Until now, in studies relating to the history of research on the passions and on expression, two methods have been utilized. The first is the one most often followed by historians of the sciences. Due to current work in the domain of the neurosciences, retrospective interest focuses on the brain. A thematic inventory that takes as its boundary the works of Darwin would make apparent the prescience of cephalocentrism over cardiocentrism, and focus on Franz Joseph Gall's doctrine, and the clinical work on lesions of the brain, with the unavoidable case of Phineas Gage and John Martyn Harlow's interpretations. The other method is the one followed by the historians of ideas. To the preceding themes must now be added pathognomy in the Renaissance, and its avatars in the classical age, the manuals about civility, the rationalization of behavior, and the classification and verification of expressions as linked to the rituals of the expressive order.

These two analyses present several points in common. First, they essentially refer to the same thematic contents. Then, they tackle the phenomena of temporal sequences according to an evolutionary schema: a succession of discourses concerning the localization of the passions and facial expressions. One recognizes continuation and its epistemological axiom, once identified by Gaston Bachelard: "Since the beginning is slow, progress is continual."[5] Finally, they fall into the same traps in linking types of research that were in fact foreign to each other, by registering them as based on observation, and by committing a series of anachronisms. One wants to discover the history of passions and of expression before Duchenne or Darwin, but one does not realize that the passions and expression did not exist then. And if they did not exist, the reason was quite simple: the brain and the face themselves did not exist. Until the middle of the nineteenth century, there existed only centers and masks. There were centers because one assigned passions to the sites of the organic life (the heart, the liver, or the ganglion nervous system). And there were masks to the extent that the social or cultural forms of expression were governed by the will. To believe that the history of passions is that of their localization, and that the history of facial expressions is that of their mastery, is to be blind to the fact that people had to lose these beliefs in order to constitute these new objects of study. To show that the passions and expression are dependent upon the animal machine, people had to renounce assigning them a site and reject the language of facial expression, for the sake of an anatomy and a physiology of affects. Duchenne, who often cited Buffon, could have applied to the face what the naturalist said of nature: "It wears only a veil, we give it a mask, we cover it with prejudices, we suppose that it acts, that it operates like we act and think."[6]

There is nothing in historians' accounts that might permit us to grasp what happened when the passions were apprehended as actions of a function and as expressive forms. In order to describe this transformation, one must abandon the idea that the history of the locating of passions is found encased in that of the brain. Also to be dropped is the idea that the history of expression is found enveloped in that of the mastery of facial expressions. Another method must be tried: the analysis of techniques, of practices, and of discourses at the moment when the passions find their point of anchoring in the play of muscles, and when their figures are sketched on the skin. Only a history of the surface enables understanding the meaning of "above" and "below" in the complexity and intricacy of this relation. Regarding which, the word "perversion" seems well suited to the system of provocation invented by Duchenne, if it is true, as Deleuze said, "that perversion implies a strange art of surfaces." As long as the anatomists possessed only the scalpel, organic structures and muscular actions remained inaccessible to analysis. As long as they deferred to the artists (or did as they did) to represent the expressive face, the details of physiognomy in movement escaped their observation. The application of new procedures of localized electrization and photography played a considerable role in this scientific reorganization. With the isolation of superficial muscles appeared a new distribution of discrete elements under the skin. With a symptomatology of passions linked to a new grammar of signs, expressive acts appeared on the surface. The elucidation of the mechanism of physiognomy in movement should be identified with the relations among these elements and with the play of their regulated functioning. Beginning with Duchenne, the gaze could follow a path that had not been opened until then: the path going from the surface to the underlying muscles, which had to be crossed in both directions to proceed to the reciprocal adjustment of muscular actions and physiognomic movements. The very possibility of a discourse on the expression of passions had to be sought in an unprecedented face-to-face. The simultaneous emergence of two faces: that of the observer who defines a new way of seeing, and that of the subject being observed who designates appearance such as it is perceived, analyzed, and dissipated within the perceptual field of the former. This is a strange face-to-face, in which Duchenne is seen to fabricate the simulacra of natural expressions. From this new scientific attitude is born his *Mécanisme de la physionomie humaine* [The Mechanism of Human Facial Expression] (1862).[7] For the first time in the history of expression, meaning no longer appears through the recognition of a silent speech, but instead is embodied in facial expression itself.

Myology

In March 1850, Duchenne sent to the National Academy of Medicine a first memorandum: *Recherches électro-physiologiques sur les muscles de la face et sur les interosseux* [Electro-Physiological Research on the Muscles of the Face and Ligaments]. Here is Auguste Bérard's review:

Gentlemen, this is not a matter of that barbaric but insufficient procedure that consisted of plunging needles into the parts that one wants to bring to electrical stimulation. M. Duchenne's way of operating is more gentle. No injections, no incisions. Humid exciters applied to the skin transmit through this membrane to the muscle underneath a galvanic irritation that the muscle obeys irresistibly. . . . Does one want to act on a large surface? A humid sponge plunged into a metal cylinder will transmit the electrical fluid to the part. Does one want to excite the delicate fractions of the muscular system—certain facial muscles, for example? One uses conical exciters covered with a humid touchwood. It is a marvelous thing to see the instrument trace the smallest radiations from the muscles! Their contraction reveals their direction and position better than the anatomist's scalpel could do! At least this is what one observes with the face, where one inevitably makes sacrifices in the preparation of the terminal portions of fibers that are going to be inserted in the internal facet of the derma. This is a new sort of anatomy to which one might apply the two words by which Haller wanted to refer to physiology: this is *animated* anatomy, *anatomia animata*; this is what Soemmerring would no doubt have called *contemplatio musculi vivi*. Today, to explain certain movements produced by some of the muscles of the face, one has to revise the arrangement of their fibers and re-write some passages in our anatomy books. . . . One reads in books devoted to these descriptions that one muscle *is continuous with another*. It

is said, for example, that the nasal pyramidal continues the frontal; that the former is continuous with some of the fibers of the muscles that drop the lower lip;
that the buccinator is continuous with the orbicular of the lips, etc. But these are
only appearances, gentlemen, and what the scalpel does not separate is distinctly
shown by the galvanic exciter. Each of these two muscles that appear to be combined in fact has its own center of action, and often between them lies a neutral
part that the galvanic exciter does not put into movement at all, as if the contractile fibers were defaulting. . . . M. Duchenne has applied his exciters to each of the
muscles that belong to the facial region. He has faithfully noted, and describes in
his memoranda, all the phenomena of contraction he has witnessed. So, what has
happened? M. Duchenne has observed new facts, but he has also seen what has
been observed before him. He has mixed in his descriptions current science with
new science.[1]

Bérard was stressing the superiority of the electric instrument over
the scalpel. Anatomical exploration establishes the independence of the
muscles. Hence the dissipation of errors linked to the fibrillary theory:
fibers are not continuations of each other. But in applying to Duchenne's
work the definition of physiology that Haller had given, *anatomia animata*,
Bérard hopelessly muddled things. The slippage of meaning from *anatomie
vivante* to *anatomie animée* meant the assimilation of myology to physiology. Because of this confusion, Bérard lost sight of the spirit of the method
invented by the Boulogne doctor. In submitting his experimental subjects
to localized electrization, he was engaging in an anatomical study that allowed him to redraw the anatomy of the face. But that the structure of
muscles could be deduced from an ensemble of phenomena linked to their
contraction did not signify that physiology was taking the initiative, for
a distinction between anatomy and physiology has to be maintained. On
the one hand, electrical excitation of the muscular substance allows one
to perceive the location, extent, and limits of each muscle. On the other
hand, electrical excitation of its motor nerve allowed its action, its specific
movement, to be recorded. Description of the various forms and functions
of muscles differs from a discourse on their usage. Moreover, there was no
reason to mix up what Duchenne himself had taken care to keep distinct.
On one side, there is the "application of direct muscular galvanization for
the study of the anatomy of forms," and on the other, there is "muscular
galvanization that is called upon to establish in an exact manner the uses
of a large number of muscles."[2] But in fact there is more there—whereas
Bérard thinks there is less, in assigning Duchenne's work a limited bearing:

his contribution to myology would seem to consist of confirming what was already known, rectifying errors, and completing the descriptions of his predecessors. But contrary to what Bérard says, Duchenne did not mix current science and new science, but rather gave an entire new foundation to the anatomy of facial muscles.

Instead of listing Duchenne's contributions to myology, we must describe a whole epistemological reorganization. No doubt most of his contemporaries shared his preoccupation. Jacques-Louis Moreau de la Sarthe, Jean-Baptiste Sarlandière, and Jean Cruveilhier also wanted to shine new light on myology and the functions of facial muscles. But their results were inseparable from their methods. As long as they were studying the muscles of a cadaver's face, the mobile attachments were being destroyed. As long as they proceeded to dissections, descriptions belonged to the framework of classical anatomy. No other organic system was as sensitive to the blade of a scalpel, which traced a watershed between the accessible and the inaccessible, between the visible (the points of origin on the bones) and the invisible (the points of mobile insertion in the skin). We indeed see why the application of localized electricity constituted a major improvement in method: anatomy on the living could respect the structure of the facial muscles. We could say that the invention of an experimental technique allowed the acquisition of new knowledge with respect to myology—but this would be inadequate. The conditions that made the reform of muscular anatomy and physiology possible are varied and belong to different spheres. First a series of difficulties of a technical nature that arose from the application of localized electrization had to be resolved. The tool could then reveal its utility because it quickly encountered all its points of application. Not only was the object of study no longer denatured, but also the motor nerves were reached at the spot where they penetrate the muscles. This means that the integration of anatomical and physiological knowledge of the cranial nerves was determining: in myology, because the electrical stimulation of the motor nerves was also a means of delimiting, with certainty, the different muscles of the face; and in physiology, because the stimulation of nerves showed their specific movements. Finally, the examination of facial pathologies had to be combined with the study of the functions of motor nerves in order to describe their tonic contractility. With Duchenne, myology was defined by the emergence of a new plane of objects that included a new plane of animation.

An Anatomy Adrift

In his *Rapport sur les progrès et la marche de la physiologie générale en France* [Report on Progress in and Pace of General Physiology in France] (1867), Claude Bernard made a judgment that may be endorsed: Antoine Laurent Lavoisier, François-Xavier Bichat, and François Magendie had given physiological studies a kind of methodological renewal that marked a "Renaissance." The physical and chemical sciences, tissue anatomy, and experimentation on living organisms had been the solid foundations of modern physiology. But anatomical physiology of facial muscles had not been affected by this movement. Several reasons account for this apparently paradoxical situation, since the least-known muscular system is the most superficial. The identification of a muscle that moves the pieces of the skeleton does not present any major difficulty. In its shortening, the location of its origin and termination give its direction, indeed its movement. One might also appreciate its action through experimenting on cadavers, by placing it either in a state of relaxation or in a state of maximum tension. By bringing closer together the points of bony insertion of the tendons (or by separating them), the displacements of pieces of the skeleton indicate the mechanism of movements. However, in the study of facial muscles, these different procedures are of no use at all. First, muscles have relations with subcutaneous tissue, but their points of mobile attachment are not identifiable: dissection destroys the terminal portions of muscular fibers. Second, in the cadaver the shape and dimension of muscles escape observation: "The body, then, is in its soft parts what amounts to an instrument, of which all the cords, instead of being stretched, are couched one under the other."[3] In short, the facial muscles had only two ways of revealing their structure and function. In the living person, during their contraction, one could detect the changes taking place in their volume and their form. Projections respond to the bodies of muscles and recesses to their insertions. One could also identify the folds that the muscles imprint upon the skin. The wrinkles of the forehead are transversal because the muscular fibers of the brow have a longitudinal direction. Here was applied the principle long since enunciated by Pierre Camper: "All the lines of the face should necessarily cut at right angles the course or direction of muscular fibers."[4]

Nevertheless, there were several attempts to update myology. In the French edition of Lavater's work, *L'Art de connaître les hommes par la physio-*

nomie (1806), Moreau added an anatomical and physiological history of facial muscles. He felt he was innovating and stressed that he was entering into new developments that were not found in the best treatises of anatomy and physiology. But the way in which the author thought of his contribution should not fool us. Moreau could do no more than his predecessors because his object of study remained the cadaver. In 1830, Sarlandière revived Camper's project of exploring the physiology of muscles in their relation to the functions of facial nerves. In the meantime Jacques Genest had published his *Exposition du système naturel des nerfs du corps humain* (1825), a translation of works by Charles Bell, followed by his own memoranda on the same subject. Charles Bell had elucidated respectively both the sensory and the motor functions of the fifth and seventh pairs of cranial nerves. At first sight, Sarlandière's program was new: no less than establishing the motor laws that govern muscular movement. "These fibers have the property to shorten under the influence of nerves and effect by this shortening the total or partial contraction of the muscles of which they are part; thus the muscle displaces the mobile parts to which it is attached."[5] But Sarlandière's project, like Moreau's, appeared within anatomical dissection of the dead. Cruveilhier was aware of the lacunae in the myology of the face, which he was the only one to have tried to remedy. Before pronouncing on the movements of muscles, he had to identify them. Since the firmness and the color of muscles facilitate their study, he gave particular attention to his choice of study subjects and to the moment of dissection. He needed vigorous individuals who had died suddenly in their prime. Nothing shrivels or pales as quickly as facial muscles. The heads of some torture victims and those of several individuals killed in the riots of May 1839 were the ones he chose to study. Until then, the muscles of the face had been studied only on their surfaces, and now their deeper facets had to be added. But as long as cutaneous insertions were being sacrificed, the study of muscles remained full of gaps.

Cruveilhier also associated the preparation of muscles with their anatomical study. Nitric acid diluted with water respects muscle fiber. By the elimination of cellular and fibrous tissues, he hardened muscle fiber and gave it a dressing that rendered it apparent. For a long time, anatomists had considered muscular fiber to be the anatomical and functional element of the muscle. This explanation going back to Aristotle had been reproduced in the seventeenth century: muscular fibers correspond to the vegetal fibers of which cords are composed. Fibrillary theory was based on the explanatory image of their function. Giovanni Alfonso Borelli established an analogy between muscular contraction and the retraction of wet rope. But Cruveilhier had a completely different point of departure: neither an explanatory image nor even what he found with the point of a scalpel, but rather what was shown by a preparation raised to the rank of an analytic procedure. This means of dissection allowed researchers to follow the continuity of fibers across intersections and the points of intersection they represented. It also allowed the muscles to conserve the form they adopt during contraction. A new fact sprang from this study: "This was *continuity* and consequently the mutual dependence of several of these muscles: thus the buccinator and the orbicular of the lips constitute one and the same muscle that we call *buccinato-labial*. Thus the pyramidal seems to be nothing other than a small piece of the frontal (*languette*); the superciliary might be considered, at least in part, a big bundle originating from the frontal and orbicular muscles; thus the square of the chin is, only in part, but in variable proportion according to the subject, the continuation of the muscle of the neck."[6] If the continuity of muscular fibers is thrown into evidence by an anatomical preparation, a muscle must be defined by its *anatomical formation* and not by its apparent form. The result is that the muscles of the face are related to the different ensembles to which they belong. These anatomical demonstrations fit within the tissue anatomy established by Bichat, and very soon they would come to dominate myology and govern its categorizations. It mattered little that Cruveilhier stressed the continuity of fibers, whereas Moreau saw "insertions" of fibers into each other, and Sarlandière saw their "fusion." At that time, anatomists agreed to define the muscular system of the face as an assemblage, whether of "apparatuses" (Moreau), "regions" (Sarlandière), or "departments" (Cruveilhier).

Since Matteo Realdo Colombo, Sylvius d'Amiens, Giovanni Domenico Santorini, Jean Riolan, and Jacques Benigne Winslow, anatomists had distinguished the small muscles of the face. Some names were derived from their purposes: the *masseter* (to masticate), the *buccinator* (to blow). Other names were derived from their shape: the orbicular, the pyramidal, the triangular, and the square. Still others were derived from either the direction of the fibers (the transversal) or their location (the occipital, the frontal, and the superciliary) and sometimes from the insertions of muscles (the large and small zygomatic). These denominations referred to distinct parts. But description of muscles was always complemented by a description of their connections—meaning that the *principle of distinction* had as a correlate the *principle of continuity*. After describing the structure of frontal muscles, for example, Winslow signaled the mingling of their fibers with those of the superciliaries, the orbiculars, and the pyramidal. The frontal muscles "cover the superciliary muscles, and are strongly adhered there by a sort of interlacing. . . . They appear to be somewhat confused with the orbicular muscles of the eyelids and the muscles of the nose." When the frontal muscles take their fixed point in the epicranial region, "their function is to lift or pull up the eyebrows, by making more or less transversal wrinkles on the skin of the forehead." But when the superciliaries, the orbiculars, and the pyramidal enter into action, due to the mingling of their fibers with those of the frontals, the latter exercise the inverse function to the preceding: "Their action is to lower the eyebrows, and to bring them closer to each other, to gather into longitudinal and longitudinally oblique wrinkles the skin that covers the bottom of the forehead above the nose and even, by irregularly transversal wrinkles, the skin that corresponds precisely with the base of the nose."[7]

By the nineteenth century, nothing had changed—except that the principle of continuity had won out over the principle of distinction, to the point that the very existence of facial muscles had been thrown into doubt. They were presumed to be secondary forms, or in Cruveilhier's expression, "borrowed muscles," borrowed from a primary space, which by coiling, superimposition, and thickening constitutes them. Muscular tissue is the common element of muscles; but it traverses them, relates them to each other, and above them, constitutes units that are concrete, massive, and anatomically indecomposable. Thus anatomists described the frontal, the superciliaries, the pyramidal, and the orbiculars as portions of one and the same very extended muscle: the fronto-orbitary apparatus. "The muscles of this apparatus hold together and are everywhere continuous, with the exception of the elevator and the eyelid: one must regard them less as particular muscles than as divisions within one and the same very extended muscle that partly covers the cranium, surrounds the eye sockets, and extends up to the superior part of the nose." The muscles of the inferior maxillary region are also described as prolongations of the muscle of the neck. The latter has close relations with all the muscles of this region, in particular the square of the chin and the muscle of the chin cleft. Similarly, the buccinator muscle and the orbicular muscle of the lips form a single muscle, the buccino-labial. To the latter must be added the ensemble of muscles of the upper lip and the corners of the mouth: the common elevator, the specific elevator, and the small zygomatic.

But shifting the preponderance of small muscles to various apparatuses did not suffice to change the terrain. To describe a fronto-orbitary region is to delimit a new anatomical unit, all of whose fixed points must still be determined. From the viewpoint of a myology called upon to sustain a discourse on muscular movements, one had to come back to the play of small muscles, even if it was thought they had only a nominal existence. They were unavoidable because they assured the pinning of each apparatus onto the bones. Therefore nineteenth-century myology was still beholden to classical myology: "We shall describe separately, and as distinct muscles, all these muscular portions, so as to better understand their uses and thus to show, in an analytical manner, the effect produced by the direction of their fibers and the combinations of their movements."[8] So much for the fronto-orbitary apparatus. But where should one situate the different points of insertions? Moreau, Sarlandière, and Cruveilhier assigned four points of attachment to this apparatus. Above, the frontal was inserted into the epicranial aponeurosis. Below, the two eyelid muscles formed the superciliary origins of the frontal since they were attached to the internal part of the superciliary arcade. As for the pyramidal, it was perceived as the pyramidal origin of the frontal since it was attached to the lower part of the nasal bone. Whether starting from the frontal muscles, as Winslow did, or from the fronto-orbitary apparatus, as Cruveilhier did, each was confronted with symmetrical and inverse requirements. Winslow had to describe the connections of the frontals with all the muscles of the upper region of the face: he came back to the fronto-orbital. Cruveilhier had to describe separately the different portions that composed this apparatus: he came back to the ensemble of muscles of the upper region of the face.

We see right away why the descriptions of the movements of muscles and skin, as one finds them in Moreau, Sarlandière, and Cruveilhier, could be no different from those of Winslow. One needed to attribute *several* movements to various portions of an apparatus. Acting in concert, they might move in different directions. The whole question was knowing where the apparatus to which these portions belonged took its fixed point. Once the connections between small muscles of the upper region of the face were described and their fixed attachments above and below were recognized, they could become *alternatively* active or passive. On the one hand, the action of the frontals presupposed the passivity of the superciliaries and the pyramidal. On the other, the action of the latter implied the passivity of the frontals. This meant that the fronto-orbital might move in two opposite directions: upward or downward. Cruveilhier was saying the same thing as Winslow: when the frontal muscle takes its fixed point from the epicranial aponeurosis, "the latter lifts the upper half of the orbicular of the eyelids, lifts the eyebrows and the skin of the base of the nose. . . . It is to this muscle that the transversal lines of the forehead are due. . . . When the frontal muscle takes its fixed point from its pyramidal and superciliary insertions, the skin of the inter-superciliary part of the forehead is lowered, and the eyebrows are carried downward."[9]

This schema of the fronto-orbital apparatus implies a perception of a muscular level molded onto the bones and covered by the under face of the skin. Through the fastening of this apparatus to the bones, it is the adherence that defines its relation to the skin. The folds seen forming on the face resulted from a linking of *surfaces*. In the first half of the nineteenth century, all anatomists were describing the anterior and posterior faces of the facial muscles: an anterior face covered by the integuments and a posterior face that rests sometimes on another muscular level and sometimes on the bony framework. For Cruveilhier, the pyramidal does not stretch from the nasal bone to inter-superciliary skin, but is interleaved between bone and the skin to which it entirely adheres. Moreover, at this time, comparative anatomy was rediscovering an educational role. Among worms, the muscular fiber becomes distinct and presents itself in the form of circular and longitudinal fibers. All locomotion is marked by either a tightening-and-dilation or a shortening-and-elongation. The movements of the facial muscles spring from these two forms of locomotion: dilation and tightening for the sphincters (labial and palpebral orbiculars), hence radiating wrinkles; and elongation and shortening for the longitudinal fibers of the apparatuses, hence the perpendicular folds in the direction of the muscular fibers. Pierre Augustin Béclard clearly indicated the nature of the relation of muscles to skin, the mechanism of contraction and its effects: "The muscles of the face, although very numerous, form a muscular plane that is not very thick . . . placed immediately under the skin. . . . Most of them gather this membrane by contracting it, and imprint wrinkles on it."[10] But the order of factors has to be inverted: it is the folds of the skin that indicate the movements of the facial muscles. The verb *froncer* ("to frown/to gather") comes from *front* ("front/forehead") in the sense of "folding." We should not underestimate the force of the image: one also frowns the eyebrows, the mouth, and the nose, like one folds a cloth by making pleats.

The essential thing to recall is that the inspection of cadavers makes apparent the intimate adherence of muscles to the skin, while the mobile attachments are hidden from sight. Curiously, the image of a muscular plane could predominate over Bichat's anatomy. Yet the latter had proposed another representation of the platysma muscles: parts that were simply fastened on one side to the bone and on the other to the internal surface of cutaneous tissue. The contraction of a muscle of the face should then move its mobile attachment on the skin by a traction movement. Bichat drew the consequence of this new explanatory image: the muscles of the face do not act together. There is no *cooperation*: "There are only muscles attached on the one side to a fixed point and on the other to a mobile point, like those of the eye and most of those of the face, which may move in an isolated manner, without necessitating a movement in other muscles."[11] Duchenne's studies might be seen as a return to Bichat's principle, beyond the theme of movements of the ensemble. For the various "apparatuses," he substituted a series of distinct muscles that have only relations of contiguity. Where previously only a relation of adherence of the superficial facet of muscles to the subcutaneous tissue had been seen, he pointed to mobile attachments. For the first time in the history of myology, a perception of the individual movements of facial muscles would be anatomically founded. Duchenne's whole enterprise was oriented toward this goal: to substitute a mosaic of distinct muscles for the muscular mask. Had not Moreau seen the apparatuses as pieces of a "muscular quilt" covered by skin and pinned to the bony framework? With touches of his electric blade, Duchenne would cut up this muscular quilt so that the small muscles could at last appear.

An Electric Scalpel

From the start, Duchenne saw what he could learn from electrical exploration. As a direct method, it allowed locating the form of muscles, the direction of their fibers, and their point of mobile attachment to the skin. Indirectly, it provoked contraction of the muscle that depended on an excited nerve. A first series of questions was being posed to this new descriptive anatomy, concerning the way to localize electricity in the muscles without provoking other movements that might cloud the analysis. To overcome the first obstacle, he had to identify the different causes that might inhibit the study of muscular actions. In the forefront of these disruptive causes was pain, since the muscles of the face were extremely sensitive. Then came the excitability of muscles, from which might result unforeseen trembling. This was because the sensitivity of muscles depended on the many nervous filaments that ran through them. Finally, how could he be really sure that the electric power had been concentrated in a muscle and that the contraction thus obtained was not the result of a reflex action? This latter was less a potential difficulty than an objection that questioned localized electrical stimulation as a research method. A second series of questions concerned the identification of points of penetration of nerves into muscles. To electrify the nerves of the face meant to localize the contraction in the muscles that they animated, in such a way as to discern their contours; it also meant studying their action by describing the movements they imprinted on the face. But this analysis presupposed knowledge of the role of muscles in the immobility of the face. This form of muscular activity seemed to constitute a more serious difficulty, because the very immobility of the face designated a muscular tension that escaped electro-physiological analysis. But integrating knowledge about the different pairs of facial nerves might serve to supplement localized electrization—provided there was a detour via pathology: the various paralyses of the facial nerve showed spontaneously the flaccidity of muscles and their effects on the modeling of the face. These deformations indicated the loss of muscle tone. They designated, by indentation, the figural posture. Treatment of paralysis of the seventh pair might also be instructive. In case of success, the doctor could see a tonic function appear in the very movement of its restoration.

The first series of questions concerned the movements that might hamper anatomical exploration. When one stimulates the muscles of a healthy subject, the pain that results provokes involuntary movements that hinder the study of muscles. How to find the proper measure that allows muscular excitation without causing the least pain? By administering small doses to attain the suitable degree of intensity, while avoiding exceeding the threshold of painful reaction. Such vivid muscular sensitivity in the facial muscles is due to the nerve filaments of the fifth pair.

So, avoid placing the exciters on points corresponding to these nervous filaments and find the insertion of the nerve that animates the muscle that one wants to make move. Sometimes, one must grope a little before finding the immersion point of the nerve in the muscle. The other difficulty is linked to muscular excitability. Here, the choice of instrument is determinative. Since the intermittences of the magneto-Faraday apparatus were not close enough together, the contractions are accompanied by a muscular trembling that perturbs the anatomical examination: hence the idea of employing a Volta-Faraday apparatus with a double current, whose intermittences were extremely rapid. Only this apparatus allows the development of continuous contractions similar to those provoked by the nervous system. Finally, in order to establish that the contractions are not due to a reflex action, it suffices to show that this function does not interfere with muscular phenomena produced by localized electrization. It was by proceeding to animal vivisections that Duchenne removed Marshall Hall's objection. Whether or not the head of a rabbit was separated from the trunk, the movements of the facial muscles were identical. But it was by experimenting on still irritable cadavers that Duchenne demonstrated the reliability of his method: "To be quite sure of producing only isolated contractions, I had to repeat on the cadaver, shortly after its death, all the experiments that I had done on the living person. For several years, no subject died at the Charity Hospital while I was visiting, on which I did not study the individual actions of the facial muscles."[12]

The second series of questions posed to Duchenne concerned the electrization of the nerves. Here Duchenne profited from knowledge of the function of each nerve of the face. Since studies by Charles Bell, it was known that the fifth pair give sensitivity to all parts of the face and preside over the movements of the masticating muscles. The seventh pair, which is exclusively motor, is distributed across all the muscles of the face. Until then, only vivisection experiments on animals had allowed study of the relationships between nerves and muscles of the face. When these relations were taken into consideration with humans, it was within the framework of treating neuralgias through electro-puncture. So it was not surprising that Duchenne registered a gap in the treatises on myology: there was no mention of the points of penetration of nerves into muscles. But what appeared to him—afterward—as an omission was in fact quite simply unrealizable. Lacking an instrument of analysis, nobody could dream of animating muscles through the excitation of nerves on a living subject. If a map of the points of immersion of nerves in muscles could now be drawn, it was because localized electrization was able to associate excitation of a nerve with the movement of the muscle that it animates. Thus it was the invention of the method that led to the determination of points of penetration of facial nerves into muscles. By this procedure, Duchenne could set the electrical power in each of the nervous filaments coming from the division of the seventh pair. Later, the anatomist Ludovic Herselfeld would design anatomical preparations that indicate the points of immersion of each of the nervous filaments into the muscles (Plate 1).

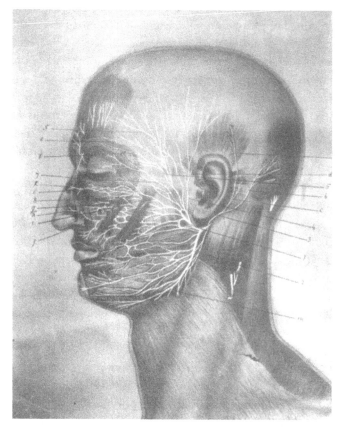

PLATE 1. Anatomical preparation of the motor nerves of the face. From Guillaume Duchenne's personal album, fig. 2. Reproduced by permission of École nationale supérieure des beaux-arts, Paris.

The electrization of the nerves facilitated study of both structures and functions. By exciting the nervous filaments that went to the muscles, it was possible to establish their anatomical independence and their specific movement. But muscles have another function: they conserve the regularity of traits through their tonicity. In this era, conditions combined to enable neuro-muscular physiology to reactivate the old notion of muscular tone, because muscle tonicity could now be distinguished from contractility. Most of the treatises in physiology or descriptive anatomy agree on defining its vital properties: contractility is inherent to muscular fibers and is independent of motor nerves. It survives the sectioning of the latter and persists for several weeks. Tonicity is a mode of contraction characterized by a continual tendency of muscles to shorten. On the one hand, there has to be intermittence of action and independence of motor nerves for the sake of contractility, and on the other, permanence of action of nerves and subordination of muscles to enervation for the sake of tonicity. The muscles are never really in a state of relaxation. No absolute rest: their action is continual. But at the start, tonic movement had been perceived as a *particular case* of muscular activity. The latter refers to the movements of the living subject in his milieu. From an epistemological standpoint, the relation of posture and muscular movement was an obligatory passage, the first problem to be resolved. This meant that the concept of tonic contractility could appear only in the extension of the study of muscular movements. One had first to elucidate the question of tonicity in more-complex muscular movements in order to tackle the much more simple question of the tonic movement of facial muscles. Duchenne had stressed the issue of chronology: his study of muscles that move the shoulder or the trunk, although published in 1855, was prior to work on facial muscles. Hence we must make a brief detour.

Galen had already described tonic movement: muscles intervened together to maintain fixed positions. But his definition of voluntary movement excluded tonic movement. In movements depending on contraction of a muscle obeying volition, its antagonist is transported. Activity consists of tension (contraction) and not the action of obeying. Passivity is inaction. Galen could not describe active immobility as a form of muscular activity, since the latter presupposes the tonic action of muscles that he considered to be inactive. Thus he had both opened and barred the way to the study of tonic movement. Winslow made a first epistemological breakthrough by effacing Galen's opposition between active and passive movements. In the movement of a limb and in the upright stance, all the muscles act together: "To move some part or to hold it in a determined situation, all the muscles that can move it cooperate in this." Hence the definition of posture tonus: "It is called tonic movement when the antagonists on this and other sides, or all the motors of a certain part, act equally and hold the part fixed between all the movements that it might make."[13] In 1778, Paul Joseph Barthez distinguished between organic sympathies and synergetic actions. He applied the term "synergy" to mechanical and idiopathic actions that preside over the execution of a function: "I refer by this word *synergy* to a concert of simultaneous or successive actions by the forces of various organs, a concert such that these actions constitute by their order (of harmony or succession) the very form of a function."[14] It was not until the nineteenth century that the term "synergy" was applied to muscular movements. But it was Duchenne who refined the schema by reactivating a central idea of Cartesian physiology. Volition commands the *dispositions* (walking, running, lifting the arm), but it has no power whatever over the synergistic mechanisms that execute its actions. The play of muscular associations, without which movements would lose their precision and posture its tone, escapes the empire of the will. "Each voluntary or instinctive movement results from the simultaneous (synergetic) contraction of a more or less great number of muscles; man enjoys the faculty of executing these movements to fulfill functions, but nature has not given him the power to localize the action of a nervous fluid in one or another muscle in a way to make them contract in isolation. . . . Therefore it is not given to us to decompose our movements and thus to analyze the specific action of our muscles."[15]

How to identify the ensemble of muscles that produce synergistic movement? Early on, Duchenne had the feeling that he was going to be able to describe what until then had escaped observation. It sufficed to substitute localized Faradization for the inspection of cadavers in order to describe the mechanism of muscular actions. The electric instrument would permit *decomposing* synergistic movements. But Duchenne soon encountered an insurmountable obstacle: not only did the action of an isolated muscle teach nothing about its functional movement, but also localized electrization altered its object of study by producing a deformation. Application of his method resulted in a failure. Duchenne could appreciate this as a problem to be solved, and in his effort to do so he would revise the whole physiology of movements. He made a new distinction between the *specific* action of a muscle and its *physiological* action. The specific or individual action is what is observed after localized electrization: for example, the deformation of the scapula that results from the artificial contraction of the deltoid. The physiological action is what is observed in the subject when he places his arm at the same degree of elevation as during the artificial contraction of his deltoid. At the instant that the arm moves away from the thorax, the acromion lifts instead of lowering; the lower angle of the scapula moves away from the median line, and the spinal edge of this bone, which is now applied against the costal wall, takes an oblique and inverse direction to what it adopted in the preceding experiment. It suffices to compare this physiological experiment with the electro-physiological experiment in order to distinguish the isolated action of the deltoid from the physiological function that it is asked to fulfill.

It remained to identify the other muscle whose assistance is necessary in the execution of the physiological movement. In order to resolve this difficulty, Duchenne had to turn toward pathology. On the one hand, localized electrization in a muscle determines its individual action, a deformation. On the other, the pathological case (as a spontaneous experiment) provides the key to decoding synergy. Locating in an injured person the same deformation leads to the identification of the atrophied muscles that in a normal state assure the movement of a limb or the maintenance of it in equilibrium by their tonic contractility. The technique of electro-muscular exploration thus found its use in revealing the absence of the muscle that contributes to the accomplishment of a physiological movement: hence the essential role of *progressive muscular atrophy*, a newly identified disease of the muscles to which Duchenne would eventually give his name. The pathological gave the analysis a new value, because it is already, by itself, the active subject that is exercised on the organism. This genuine principle of division among organs, under the effect of morbid alterations, could operate by elimination. Basically, the pathological case indicates muscles whose synergistic action is hidden from view. It suffices to have an arm lifted by a man suffering from an atrophy of the major serrated, detectable by electric exploration, to see appear the same deformity that is provoked by making the deltoid of a healthy subject contract in isolation. In comparing this electro-muscular experiment with the clinical case, Duchenne concluded that the synergistic action of the major serrated allows the deltoid to fulfill its physiological function. In this study of physiological movements, tonic movement indeed appeared as a particular case of muscular synergies: "The muscles of the shoulder are not only designed to imprint their movements upon it; we know they are also kinds of active ligaments that maintain the scapular and humerus in their respective normal positions, setting aside all voluntary contractions. This force, which allows the motor muscles of the shoulder to act on it like active ligaments, even when these muscles do not contract under the influence of the will, is tonic contractility."[16]

We immediately see why the elucidation of the tonic movements of the muscles of the face did not present any major difficulty. They, unlike the muscles that preside over ensemble movements, have only one individual action. Their tonic contractility, like their contraction, is also their function—and failed action is its index. Of course, Charles Bell had demonstrated the motor action of the facial nerve through vivisection experiments, but his section abolished the movements of the face. And it is also true that the general symptomatology of the paralysis of the facial nerve had been traced by the Englishman and by Bérard. But as long as the different facial muscles had not been identified, one could not dream of exploring their tonic function. Only Duchenne could integrate pathological observation as a way of indicating the loss of muscular tonicity. He applied to the facial muscles the same distinction that he had just made between cerebral paralysis and spinal paralysis. In facial paralysis with a cerebral cause, contractility is intact. On the other hand, in symptomatic facial hemiplegia of a lesion of the facial nerve, contractility is abolished in paralyzed muscles. On the basis of this sign that authorizes a differential diagnosis, the tonicity of the muscles of the face is put into evidence in facial hemiplegias. Paralysis of the seventh pair allowed the study of the influence of the tonicity of each muscle for each facial feature—for example, in the orbicular of the lips, the soft parts of the paralyzed half are drawn toward the healthy half. This deviation of the angles of the mouth is due to the tonicity that exists on one side but has disappeared on the other side. But electrotherapy is also instructive. The restoration of tonic force, which precedes the return of muscular movement, redesigns the regularity of features. Thus Duchenne "saw" the effects of the tonic function reappear.

Curiously, in reaching beyond his own or any classical era, Duchenne was reviving the principles established by the founder of modern anatomy. Andreas Vesalius had rejected any reference to animals as a possible substitute for man. He had established as a methodological imperative the idea that the human structure could be observed only on man. Vesalius was the first to become aware of the possibilities offered by the dissection of cadavers within the framework of anatomy. The human body is the only faithful document on the fabric of the body. Duchenne pushed as far as possible Vesalius's principle that one must study what is human on man himself, but he had given it significance and range without precedent. In stressing the unsuitability of the human cadaver for study of internal morphology, descriptive anatomy effected an ultimate disarticulation, heavy with consequences. Before subordinating the form to the movement, the structure of muscles had to be apprehended in an analysis that respected their integrity. An "animated myology" had to be invented. Georges Canguilhem had stressed Vesalius's singular position—"the unification in a single man of the three personages in the ancient anatomy lessons: *magister, demonstrator, ostentator.*"[17] Duchenne, like Vesalius in his day, assumed the triple function that allowed him to revitalize myology. In the midst of the nineteenth century, with the same independence of mind as Vesalius, Duchenne taught, demonstrated, and showed. The doctor of Boulogne, like Vesalius, closely linked theory and practice. Publishing the aptly titled *De l'électrisation localisée* (1855), he devoted almost two hundred pages to a minute description of the instruments and techniques of electrical experimentation. According to legend, Vesalius is said to have dissected a living man who revived under the scalpel. He had died during a shipwreck when he was on a pilgrimage to expiate his sin. This story summarizes the profound ambivalence of what an anatomy worthy of the name might be, but unworthy of the one who would practice it. Three centuries later, the myth became reality: Duchenne invented vivisection on a human, without sectioning: "It was, so to speak, the *living anatomy* practiced on animals by the ancients, and realized for the first time on a living person, without a bloody operation, with inoffensive procedures, due to the advances in my method of localized Faradization."[18]

Dissection of the Living

The paper of 1850 went to the core of Cruveilhier's doctrine to stress its contradictions and undermine its foundations. If the anatomical unity of a muscle was defined as an apparatus, then not only was the description of small muscles unjustified, but their denomination had to be eliminated from the nomenclature. In this form a dilemma arose regarding what until then had been considered as compatible: either a muscle is assimilated to an apparatus, or else it is identified with a portion of this apparatus. And any delimiting of a portion conferred independence on it, by which it acceded to the status of an anatomically distinct muscle. This anatomical recomposition, which was assuming the appearance of a fragmentation of the muscular mask, was due to a new kind of anatomical and physiological experiment (Plate 2). Apparently a return to the description of the small muscles, as already found in Riolan, Bernhard Siegfried Albinus, and

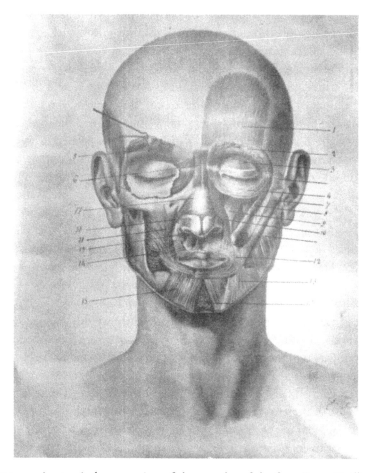

PLATE 2. Anatomical preparation of the muscles of the face. From Guillaume Duchenne's personal album, fig. 3. Reproduced by permission of École nationale supérieure des beaux-arts, Paris.

Winslow, this new description differed on one essential point: study of connections among them had completely disappeared. In truth, Duchenne had revived Bichat's principle and its corollary—the individual actions of the orofacial muscles. But thanks to electro-muscular exploration, this principle now demonstrated its fecundity: "I hope to prove soon with the help of galvanism that the muscles of the face, which according to the fine studies of M. Cruveilhier would appear to be placed in reciprocal dependence by the apparent continuity of their fibers, are in reality perfectly isolated and independent one from another."[19]

How to identify the various muscles of the face? The first stage consisted of locating the neutral points separating two muscles. Since the latter are contractile throughout their continuity, the absence of a contraction designates their terminals on the internal part of the skin. At the level of the inter-superciliary region, the exciter indicates a furrow: an immobile line where muscular fiber does not exist. On each side of this line are found points of mobile attachment to the skin, which are sufficient in relating the points of origin. The skin is pulled downward by the pyramidal of the nose. This muscle is formed of vertical fibers whose lower extremity is attached to the bone of the nose and whose upper end is in the skin, at the level of the inter-superciliary space. But when the exciter is applied above the neutral line traced by the transversal inter-superciliary groove, the skin is entrained upward by the frontal. This muscle takes its mobile point on the skin, below, and its fixed point on the epicranial aponeurosis. The pyramidal and the frontal are independent and antagonistic muscles. The second operation consists of identifying the shape of a muscle by moving the exciter over wet skin. From the instant the exciter finds itself outside the limits of the muscle, movement no longer takes place, or else it is modified. Duchenne confirmed in this way the independence of the pyramidal and the frontal: the electrical excitation of the fibers of one does not pass into the other. He also showed the independence of the frontal and the superciliary portion of the superior palpebral orbicular. By the same method, finally, he established the independence of muscles of the lower part of the face. The continuity of fibers of the muscle of the neck in the triangular of the lips and the square of the chin had persuaded most anatomists to consider the latter as prolongations of the former. In reality, a clear boundary separates the muscle of the neck from the triangular and from the square of the chin.

A description of the muscular actions and displacements of the skin that resulted from them was becoming possible. But it presupposed elucidation of the tonic function of the muscles. Active immobility designates the degree zero of movement, on the basis of which the various modifications of the physiognomy might be perceived. Hence the third operation: examination of the deformations of features after the loss of tonicity, and by means of electrotherapy, the recording of the return to tonicity from which results the restoration of features. The loss of tonicity of the zygomatic major occasions the fall of the corner of the lips, the effacement of the lower part of the naso-labial groove, and a lowering of the cheek that entrains the lower eyelid downward. The Faradization of the paralyzed muscle gradually revives its tonicity: the corner of the lips lifts through the tonic retreat of the grand zygomatic, the cheekbone goes back up, and the lower eyelid sticks to the superior palpebral. The disappearance and the return of tonicity placed Duchenne before the ensemble of the effects of the tonic action of the zygomatic major. At rest, tonicity assured the rectitude of the corner of the lips, the regularity of the naso-labial groove, and the swelling of the cheek that maintains the lower eyelid in a natural relation with the upper lid. Previously, all anatomists had associated a single physiognomic movement with the action of the zygomatic major: the corner of the lips is lifted up and out. In reality, the isolated contraction of this muscle determines several modifications: the corner of the lips is drawn upward and outward; the line that separates the lips describes a curve with upper concavity; the naso-labial groove is less curved and farther away from the median line; the raised cheek becomes more prominent; and the skin, pushed up from below, is covered with radiating lines surrounding the eyelids.

While Duchenne as of 1850 was revising anatomy and the physiology of facial muscles, he still had not yet elucidated the relations between muscular actions and physiognomic movements. No doubt the pyramidal was no longer confused with the frontal. But when Duchenne spoke of the physiognomic movements that the muscles determined, he invoked the distinction between expansive and oppressive passions. The contraction of the pyramidal darkens the physiognomy, while that of the frontal lights it up. Duchenne was well aware that his results were insufficient: "This first work, despite the reception given it at the Academy and in the medical press, was still and could only be a rough draft. The electro-physiological data that I had observed did not completely account for the physiological movements of the eyebrow."[20] Not until 1857 did Duchenne resolve all the problems. Then and only then could a first link be established between anatomo-physiological analysis of muscles and their expressive function. The paper he presented in competition for the Volta Prize bore direct witness to this. In effect, this unpublished paper of March 1857, titled "Considérations générales sur la mécanique de la physionomie" [General Considerations on the Mechanics of Physiognomy], would form the first part of the eventual *Mécanisme de la physionomie humaine* of 1862. But this paper is incomplete, ending with the words "To be copied with deductions applicable to the fine arts and a chapter titled *movements of the eyebrow*" (p. 31). In fact, this chapter relating to eyebrow movements was later published in the *Physiologie des mouvements* (1867) under the title "Recherches anatomiques et expérimentales sur les muscles du sourcil" (pp. 815–827). The editors, who seemed ignorant that the chapter did not figure in the dossier for the Volta Prize, nevertheless confirmed its provenance and the date when it had been written: "This note is an extract from a paper addressed by M. Duchenne (of Boulogne) to the competition opened at the Academy of Sciences in 1856 by His Majesty Napoleon III, on the application of electricity to medicine and to the industrial arts" (p. 815). Thus a period of latency separates the first anatomo-physiological analysis (1850) from the analysis of the physiological mechanisms of expression (1857).

This was the time passed before Duchenne encountered the experimental subject who would allow him to resolve all problems relating to the anatomy and physiology of the fronto-orbital region. Until then, the study of muscles ran up against a difficulty: the reactions of experimental subjects frustrated the electrical exploration. Exciting the sensitive nerves that run to and fro through the subcutaneous layer of the face was excluding a precise delimitation of the muscles of the upper part of the face. This was because electric exploration provoked, if not a painful reaction, then at least involuntary movements that barred a detailed perception of structures and movements. This complication could not be avoided except by making experiments on an almost insensitive subject. Hence the important role played by his new experimental subject: "This man showed a very favorable condition that I had not recognized in other subjects. . . . This subject had reduced sensation. He was suffering from a complicated anesthetic condition of the face. I was able to experiment on his face without causing him pain."[21] So we may be sure that the year 1856 was decisive; if Duchenne had examined this clinical case before 1855, he would have reported it in the first edition of *De l'électrisation localisée*. Duchenne gives a description of this case for the first time in a "Note on Functional Spasm and Functional Muscular Paralysis" published in the *Bulletin général de thérapeutique* in 1860: "A cobbler felt the same contractures in the right rotators of the head, and in some muscles of the shoulder right and the face, as soon as he started to work" (p. 146). This observation was taken up in the second edition of *De l'électrisation localisée* (1861, observation 224, p. 930) and of course in the *Mécanisme de la physiognomie humaine* of 1862. It also figures in the third edition of *De l'électrisation localisée* of 1872, where Duchenne gives a precious piece of information: "He worked at least 12 hours out of 24, which had provoked his functional spasms. I had him enter the Charity Hospital in the service of M. Briquet, where I tried to treat him by Faradization of the antagonists of the contractured muscles. It was during this treatment that I did my electro-physiological experiments on his face."[22] In effect, the chapter of 1857 dealing with eyebrow movement is the result of experiments done on the shoemaker.

We see right away why these experiments were crucial. Duchenne could proceed to electrization of nerves and call upon touching. At the instant when a muscle contracts, all the muscular fibers enter simultaneously into action. It suffices to move the finger over the skin at this place to delimit the space where the contraction makes itself felt. By digital examination, one felt a vibrating movement of the muscle everywhere that the muscular fibers composing it extended. In 1850, the independence of the pyramidal and the frontal had been accepted, but for other muscles, things were not so clear. Duchenne thought, as did Cruveilhier, that the superior part of the palpebral orbicular and the superciliary constituted one and the same muscle: "This electro-muscular study led me to see the superciliary just as a dependence of the orbicular, of which it constituted one of the attachments."[23] In 1857 new segmentations appeared. The extra-palpebral orbicular was distinct from the superciliary. If you placed the exciter at the level of the motor nerve of the superior extra-palpebral, then you felt the vibratory trembling in the upper half of the superciliary arcade, except for its internal part. If you placed the exciter at the point where the motor filament of the superciliary muscle is subcutaneous, you felt vibratory tremblings in a continuous length corresponding to the internal part of the superciliary arcade. "It seems to me these experiments demonstrate that there really exists a line of demarcation between the fibers of these different muscles that in appearance are continuations of each other. If these points of separation did not exist—if, as has been claimed, the fibers of the frontal, the superciliary and the superior half of the orbicular portion were continuations of each other—then the muscular contraction would be felt throughout the length of these very short muscular fibers."[24]

In 1857, Duchenne described a series of four movements: the eyebrow is entrained in various directions by four special muscles. Two of these muscles raise or lower it *en masse*: the frontal lifts it and the upper half of the extra-palpebral lowers it. The two others raise or lower only its internal extremity: the pyramidal lowers it and the superciliary raises it. Thus he was analyzing relative movements. On the one side, the upper extra-palpebral orbicular lowers the eyebrow; it is congeneric with the superciliary for the tail end of the eyebrow (its external two thirds), and its antagonist for the head of the eyebrow (its internal third). On the other, the superciliary lifts the head of the eyebrow; it is the antagonist of the pyramidal and the superior orbicular for this part. But it is also the antagonist of the frontal for the tail of the eyebrow, and is congeneric for the head. Duchenne's perception is structurally different from that of his contemporaries. The modifications that he describes define a matrix (*combinatoire*). This phenomenological and technical analysis is opposed to the description of movements of an apparatus, just as an analytical perception contrasts with a vision of the ensemble. Action of the frontals: the frontal lines extend over the whole width of the brow; on each side, they describe curves of inferior concavity that, in gathering on the median line, form a new curve of superior concavity. Action of the superior palpebral orbicular: the eyebrow is lowered *en masse* and the frontal lines disappear; it becomes rectilinear and executes a movement of corrugation. Action of the pyramidal: the skin of the inter-superciliary space is drawn downward; at the level of the head of the eyebrow there appears a transversal furrow. The action of the superciliary is the most complex. By its contraction, the head of the eyebrow is inflated; the eyebrow has become oblique from top to bottom and from inside to outside; it describes a sinuous line composed of two curves, one internal of superior concavity and the other external of inferior concavity. Moreover, it has developed several transversal cutaneous folds on the median part of the forehead, and outside these folds the skin is stretched above the internal half of the eyebrow.

It would not be false to say that with electro-muscular exploration, physiology won out over topography. But Cruveilhier, who opted for the topographical order, said the same thing as Winslow, who was following the physiological order. The essential thing lay elsewhere. The line of separation was not between two techniques of description but between two methods of analysis: an anatomy that proceeds to inspection of cadavers and an anatomy that explores the muscles and their movement on the living. The first precludes a discourse on the usage of the small muscles, since they are perceived as portions of an apparatus. The second identifies the small facial muscles and registers for each of them its fixed and mobile points of attachment. For the first time in the history of the face, a muscular physiology accounted for the diversity of what is modeled. Duchenne's research on the anatomo-physiology of the facial muscles was quickly tied up—in any case, well before a physiology of the movements of limbs on the trunk. At the time, the best treatises of myology were giving the impression that muscular physiology had already acquired the stature of a rigorous discipline. In 1843, Cruveilhier wrote: "The physiological order of Vesalius and Winslow has been reproached for being defective, given that the uses of a great number of muscles are still undetermined. This objection might have been serious in the time of the great anatomists, but today, science is too advanced for one to be mistaken about the opinion on the action of muscles."[25] Cruveilhier, of course, was speaking of the actions of muscles that move pieces of the skeleton and not of orofacial muscles: the former seemed well known, the latter escaped analysis. Starting with Duchenne, the situation was the inverse: the anatomical and functional independence of facial muscles was brought to light, whereas a new opacity enveloped the physiology of synergetic movements. To proceed to study the mechanics of ensemble movements, Duchenne had to combine clinical observation with his electro-muscular analysis. To the extent that the physiology of movements was mobilizing the muscular pathology of all the muscles, it could only be a long, drawn-out enterprise. On the other hand, the method of electrization was being applied without difficulty to facial muscles: they are superficial and each has only one specific movement. If the function of facial muscles had been rapidly elucidated, it was because localized electrization was the precise and correct *method of analysis*. It is true that Duchenne had encountered difficulties, but these were surmounted as of 1857.

2

Machinery

In the introduction to *The Expression of the Emotions in Man and Animals*, Charles Darwin gave a brief overview of studies of facial expressions:

Many works have been written on expression, but a greater number on physiognomy—that is, on the recognition of a character through the study of the permanent form of the features. With this latter subject I am not here concerned. The older treatises, which I have consulted, have been of little or no service to me. . . . The *Discours*, delivered 1774–1782, by the well-known Dutch anatomist Camper, can hardly be considered as having made any marked advance in the subject. . . . Sir Charles Bell . . . does not try to explain why different muscles are brought into action under different emotions. . . . In 1807 M. Moreau edited an edition of Lavater on physiognomy, in which he incorporated several of his own essays. . . . He throws, however, very little light on the philosophy of the subject. . . . There is but a slight, if any, advance in the philosophy of the subject, beyond that reached by the painter Le Brun who, in 1667, in describing the expression of fright, says: "The eyebrow, which is lowered on one side and raised on the other, gives the impression that the raised part wants to attach itself to the brain, to protect it from the evil that the soul perceives.". . . I have thought the foregoing sentences worth quoting, as specimens of the surprising nonsense which has been written on the subject. . . . In 1862, Dr. Duchenne published two editions, in folio and octavo, of his *Mécanisme de la Physionomie Humaine*, in which he analyzes by means of electricity, and illustrates by magnificent photographs, the movements of the facial muscles. . . . As it is known that he was eminently successful in elucidating the physiology of the muscles of the hand by the aid of electricity, it is probable that he is generally in the right about the muscles of the face. In my opinion,

Dr. Duchenne has greatly advanced the subject by his treatment of it. No one has more carefully studied the contraction of each separate muscle, and the consequent furrows produced on the skin. Notwithstanding Dr. Duchenne's great experience, he for a long time fancied, as he states, that several muscles contracted under certain emotions, whereas he later convinced himself that the movement was confined to a single muscle. . . . Dr. Duchenne galvanized, as we have already seen, certain muscles in the face of an old man, whose skin was little sensitive, and thus produced various expressions, which were photographed on a large scale. It fortunately occurred to me to show several of the best plates, without a word of explanation, to above twenty educated persons of various ages and both sexes. . . . Several of the expressions were instantly recognized by almost everyone. . . . On the other hand, the most widely different judgments were pronounced in regard to some of them. This exhibition was of use in another way, by convincing me how easily we may be misguided by our imagination.[1]

Here Darwin reviews authors whose views are no longer acceptable, from Le Brun to Charles Bell, without forgetting Lavater, Camper, and Moreau, offering a teratology of knowledge that is brusquely pushed to the margins of a new discipline. A whole philosophy of the passions has *toppled into the false*. But make no mistake—Darwin's "historical survey" has nothing to do with a history of expression; it is the judgment of history and not the history of judgment: seen from inside a new discipline, the past is disqualified. In contrast, Duchenne's works mark a transformation of the field of psychology. In saying that Duchenne is "in the right," Darwin situates him where he places himself: on the new horizon of a scientific psychology. With reason, Darwin insists on the heuristic role of the images: "[Duchenne] analyzes by means of electricity, and illustrates by magnificent photographs, the movements of the facial muscles." He also stresses the difficulties that the doctor of Boulogne had confronted. At the start, Duchenne allowed himself to be trapped by an illusion. He thought that the emotions were expressed by synergetic movements: "Notwithstanding Dr. Duchenne's great experience, he for a long time fancied, as he states, that several muscles contracted under certain emotions, whereas he later convinced himself that the movement was confined to a single muscle." On that basis, Duchenne cast doubt on Camper's principle that each fold of the face is the effect of the contraction of the underlying muscle: "No one has more carefully studied the contraction of each separate muscle, and the consequent furrows produced on the skin." Finally, Darwin associates the photographs with the pedagogic project of their author. The images take their full significance from the texts that accompany them, and their truth appears only to the eyes of informed viewers. The explanations eliminate many errors: "This exhibition was of use in another way, by convincing me how easily we may be misguided by our imagination."

We should pause a moment on what Duchenne and Darwin had definitively rejected: physiognomy and pathognomy. But we need not go back as far as the Renaissance. Descartes, in his *Traité des passions de l'âme* [Treatise on the Passions of the Soul, or The Passions of the Soul] (1649), made a physiological study of the signs of passions that had undermined this whole tradition. Moreover, in excluding a theory of expression, he blocked in advance Marin Cureau de La Chambre's book *L'Art de connaître les hommes* [The Art of Knowing Men] (1659), and Le Brun's *Conférence sur l'expression générale et particulière* [Lecture on General and Particular Expression] (1668). Later in the eighteenth century, when historians agree there was a resurgence of the old pathognomy, it would be necessary to find a new way of linking the interplay between physiognomy and the philosophy of the natural sign. Without integrating Lavater into this history, one could not comprehend the very first encounters in the physiologies of Moreau, Sarlandière, and Cruveilhier with the problems of expression. Nor could one comprehend the failure of this new pathognomy that took over from that of Zurich's pastor, which presented the ensemble of obstacles that Duchenne had to overcome in order to found the mechanism of passions. In order for pathognomy to prove problem-free and to enable the passions to be regarded otherwise, it had to constitute new objects of study. But the revision of anatomy and muscular physiology was a necessary yet not sufficient condition. It was the photographing of electro-physiological experiments that opened the way to fresh research on expression. Representation by means of optics went beyond the simple recording of an event; it was the objectivation of a perception and the point of departure for physiognomic analysis. Photography captured the contours, a surface effect that had the status of object of study: the presentation of a facial expression as it appeared. The mechanism of physiognomy is the analysis of actual movement in its relation with the image as appearance. New objects would give themselves to psychology, to the extent that the knowing subject starts to function according to a new model. What is going to change is the whole ensemble, the relation of expression to this gaze to which it offers itself, and which at the same time it constitutes. The rec-

tification of the immediate data of perception is inseparable from the anatomical analysis of passions. This new structure of visibility presupposes the
discovery of optical illusions, the integration of individual differences into
the study of expressive forms, as well as knowledgeable pedagogy.

The Limits of Pathognomy

In *Traité des passions de l'âme*, Descartes encountered the question of
the corporeal modifications that accompany the passions. These transformations are presented in the form of signs: "The principal of these signs
are the actions of the eyes and face, changes of color, tremors, languor,
swooning, laughter, tears, groans and sighs." But these signs are differen-
tiated as much by their causes as by their signification. Some originate in
the heart and their manifestations are involuntary: they must be symptoms. Others are determined by the soul and their manifestations are voluntary; at first sight, they have the aspect of indices. Signs that are related
to symptoms occupy a central place: of all that is visible, the changes in
skin color, for example, are closest to the essential and are among the principal effects of movements of the heart. Joy makes one flush, because the
warmer and subtler blood flows more quickly; the face appears swollen,
laughing. Inversely, sadness narrows the coronary orifices, and the colder
and thicker blood flows more slowly; the face appears pale, thinner. "We
cannot so easily prevent ourselves from flushing or becoming pale when
some passion disposes us to do so, because these changes proceed more
immediately from the heart, which may be called the source of the passions, inasmuch as it prepares the blood and the spirits for producing
them." The other indexical signs are the actions of the eyes and the face
that depend on nerves and muscles. Right away, they present themselves
as indices that enable recognition, since facial actions that accompany the
passions designate kinds of looks: "It is true that there are some that are
remarkable enough, as are the seams in the forehead which come in anger,
and certain movements of nose and lips in indignation and in scorn." But
one would be wrong to rely on them: "They do not so much appear natural as voluntary. And generally speaking, all actions, whether of face or
eyes, may be changed by the soul when, desiring to hide a passion, it vig-
orously calls up the image of a contrary one: so that we may make use of

these actions as well in dissimulating our passions as in evidencing them." To feign one passion in order to hide another means to use the same signs by which it happens that a passion is declared. Mimicry is not the *expression* of a passion, but the sign of a language whose meaning, although implicit, appears clearly. "There is no passion that is not evidenced by some particular action of the eyes that is so manifest in certain emotions that even the stupidest servants can remark by the eye of their master if he is or is not angry with them."[2] Here Descartes alludes to that kind of abrupt and manifest anger that may be easily calmed, since the master is behaving as one does when not wanting to take revenge other than by facial expressions. In short, Descartes opposed natural movements to intentional movements, like cardiopathy to mimology.

By this distinction, he establishes the relation of facial expressions to the field of passions. But this relation is problematic. At the very moment when Descartes was integrating expressions as the external signs of passions, he was showing that one question remained suspended. The stupidest servant, or the most intelligent, will never know if the master is declaring his anger or feigning his anger. Voluntary and not natural, expressions still signify passions, eloquent signs that are as unreliable as the will that can alter them. The first refutation of physiognomy is that mimicry of expressions, like speech, may be bearer of truth or falsehood. But not all facial expressions are voluntary: for example, the laughing that seems to come from joy or weeping. Is it sufficient for a look to be natural for it to become an authentic index? When looks are so, Descartes identifies them with symptoms. Among all of these symptoms, there is not one single expression where total certainty may be attached to *one* passion. Signs are sometimes undecipherable: there are people who make almost the same expression when they cry as others when they laugh. Signs lack specificity: they might be caused by different passions. Languor is occasioned by love and by desire, but it might also be caused by hatred, sadness, and joy when they are violent. Signs are unpredictable: the sign that customarily appears in one passion might give way to a sign of an opposite passion. Joy makes one turn red and sadness makes one go pallid, but one may flush while being sad. This is because sadness is combined with other passions, like love or desire, hatred or shame, that make one flush. Signs are situational: laughter does not always accompany joy, nor crying sadness. Signs are uncertain, taking into account the causes and mechanisms that produce them. Tremors might be as much provoked by sadness, fear, or cold air as by desire, anger, or drunkenness: in the last instance, too many spirits flow to the brain and in the first, too few. Similarly, laughter may proceed from joy, admiration, hatred, aversion, or hunger. In joy and admiration, it is the effect of an excess of blood consecutive to the dilation of the cardiac orifices; in aversion, the rate sends a subtle fluid that augments the rarefaction of the blood; in hunger, the first digestive juices that pass from the stomach toward the heart provoke the dilation of the lungs. In sum, symptoms are not pathognomic signs. This is a second refutation of physiognomy: all the signs of passions drawn from symptoms are ambiguous.

Moreover, symptoms (as natural movements that accompany passions) are signs that emerge to the surface area. One would be wrong to take them as a system of signs similar to the words or gestures of deaf-mutes when they express themselves. To the extent that the signs evidencing passions proceed from the sole disposition of the bodily machine, they have the same nature as the movements one observes in animals. "We ought not to confound speech with natural movements which betray passions and may be imitated by machines as well as be manifested by animals."[3] The third refutation of physiognomy is that physiological determinism is sufficient to dissipate the illusion of a language-like structure of external signs. Hence it is impossible to apprehend the nature of the soul by means of the external signs that accompany passions. Our thoughts are actions that come from the soul and are accomplished in it. By contrast, our passions come from the body and are received in the soul. Language belongs to thought: there is no *relation of expression* between soul and body. When it happens that a person makes himself understood through his expressions, this is by a silent language that might be sincere or false. The face is not the mirror of the soul and it does not speak like an oracle. In 1659, Cureau de La Chambre, author of *Les charactères des passions* (1640–1659), to whom Descartes had presented a copy of his passion treatise, wrote: "Nature has given man a voice and a language not only to be the interpreters of his thought, but in the suspicion that he might abuse them; she also made his forehead and his eyes speak in order to give them the lie when they are not faithful. In a word, she has spread his whole soul on his outside."[4] Pierre Séguier's doctor seems behind the times. He relies on a doctrine whose foundations had been destroyed by Descartes. Cureau de La Chambre was referring to a sensitive appetite, itself subdivided into a concupiscent appetite and an irascible appetite. To the former he attributed two pairs: love and hate, pleasure and pain. To the latter: courage and fear, constancy and consternation. As for mixed passions, they are composite or simple, like the anger that depends either on pain or on courage. He identified appetite with the seat of the passions, and defines them as movements of the soul—real and immanent movements by which the soul does not change its place, although its parts take diverse situations "in the same way as the water enclosed in a vase might be agitated in its parts without changing place."[5] When the soul moves, it communicates its movement to a body that alters and changes. This is why the soul's *actions* appear on the face and there imprint their mark. Later, when Le Brun gave his *Conférence sur l'expression générale et particulière*, he would define passion as a movement of the soul that resides in its sensitive part, thereby following Cureau de La Chambre and remaining on the terrain of scholastic philosophy.

Historians say that it was not until the second half of the eighteenth century that we witness the renaissance of physiognomy. But the studies of Lavater and Camper do not follow in the same line as those of their predecessors: under the same term, we must perceive a different project; under this phenomenon of an apparent return, we must register a transformation. Cureau de La Chambre and Le Brun had been inscribing the expression of passions within the framework of a physiological determinism: physiognomic traits are the consequence of movements of the soul. However, Lavater and Camper lodged the relation of invisible to visible in the realm of semiology alone. Expressions were just the signs of passions. Le Brun was sketching in order to grasp the relations of a human physiognomy with that of animals; his research remained within the tradition of Renaissance bestiaries. Lavater and Camper are sketching in order to know; a portrait, a silhouette, and the form of a cranium are distribution points for an ensemble of signs. Graphic representation, or the description of a prototype, is already *physiognomic writing*. The last difference derives directly from the preceding ones: it is on the basis of physical resemblances between man and beasts that Le Brun wondered about the signs that marked natural inclinations. Quite other is the point of departure of Lavater and Camper: not only the catalog of emblematic animals, but the physical traits that represent character. One would be wrong to see the emergence of craniometry, or of measuring the forehead, as an instrumentalization of observation. The measure of the facial angle and the curve of a forehead are not signs, but they become so from the moment when character deposits—or rather, transposes—its qualities onto the face. In the eighteenth century, the mutation of the gaze was linked not to measurement but to the formation of the physiognomic method: a constitution of a domain of visibility where the gaze perceives signs right from the start. In their material reality, signs are identified with their morphological support. But the analysis of signs is the decoding of interior qualities. This means that the parts of a silhouette can receive their meaning only from an act of knowledge that has already transformed them into signs: "Each of these parts, considered in itself, is a character, a syllable, a word; often a judgment, an entire discourse of nature, always truthful."[6] How does this operation that transforms a facial feature into a signifying element, which signifies judgment as its immediate truth, take place? By a procedure that deposits in the field of perception a language of forms that the observer has only to read in order to re-transcribe it into his own discourse. When he sees a face, the physiognomist reads it like a text, or hears it as a language.

Physiognomy is nothing other than a semiology coupled with a hermeneutics. The apprehending of intellectual, moral, and animal faculties is filtered through the theory of signs. The sign is an element taken from the body and constituted as a sign by knowledge. The physiognomist must observe features, lines, and contours of the face because they are natural signs. "He sees and he is obliged to see what is before his eyes. The object that presents itself is the image of what he is not drawn to perceive."[7] What offers itself to his gaze is a reflection that designates what it reflects. The different parts of the face are mirrors that present images of invisible qualities: the forehead down to the eyebrows is a mirror of the intelligence; the nose and cheeks are a mirror of the moral and sensitive life; the mouth and chin are a mirror of the animal life. But the structure of physiognomical perception presupposes that two elements and the relation between them are already known: a physical feature is the representation of, and the substitute for, an interior quality. That which assures that the relation of the sign to the thing that it signifies is not arbitrary, is the identification of indices. The physiognomist possesses the talent to know the interior of a person from his exterior; he perceives by certain indices that do not immediately strike the senses. The value of the sign rests on the relation, given and perceived, between the index and what is indicated. The index is a marker that does not deceive, taking into account its frequency.

For example, eyes full of fire, a gaze as prompt as lightning, and a pene-trating mind are often found together. This structure of indication poses a relation between the physical and the moral. But in order for each sign to carry what is signified, it was necessary for it to be given the same knowl-edge as it signifies. As Lavater suggests, a physical feature would never have become the sign of a character trait if it were not detected at the moment when the latter is perceived. Hence the decisive role of experience—that is to say, of analysis and comparison. One must study extreme charac-ters: on the one hand, the traits of excessive goodness and, on the other, those of the darkest malice: a poet full of imagination and heat, an apa-thetic spirit whom nothing can move; a born imbecile and a man of talent. Lavater thus maintains that we judge the qualities of a person not by his actions but by his face. But he did not perceive that with the description of portraits, he was giving in advance the qualities that he was claiming to deduce from the representation.

What then is the status of the physiognomy in movement? At first sight, Lavater has clearly distinguished physiognomy from pathognomy: "Physiognomony [*sic*] explains the signs of faculties of character. Patho-gnomony [*sic*] is the interpretation of passions, or the science that treats signs of passions." In reality, Lavater sees physiognomy in movement only as forms of the expression of character: "The former envisages character in a state of rest, the other examines it when it is in action."[8] Several conse-quences follow from the subordination of pathognomy to physiognomy. The expression of passions is not envisaged as the distinct object of study of character. It is not physiognomical movements that retain Lavater's attention but what of the original character is manifested in them. It might happen that pathognomy confirms the value of a sign. To be sure that two characters resemble each other, one must observe the moment at which they are put into movement. If the line arising from the movement of muscles is the same in the two faces, then the conformity between spir-its cannot be in doubt. But, more often, physiognomical play blurs the study of character. To the extent that the original dispositions are inscribed in the solid parts, physiognomy in movement hides them. Each of us is supposed to have a primitive physiognomy whose origin and essence are divine. It reestablishes itself in the calm of death, just as troubled waters clear when they are no longer agitated. Hence the marginal position of pathognomy in Lavater's work. It could not be otherwise, since physiognomy runs through a space of visible and immutable variables, which is incompatible with the functioning of organs. At its interior boundary, the surface of the face adheres to muscular planes that are themselves molded onto the bones of the head. Craniopathy, which obeys the same rules as physiognomy, excludes myopathy as essential, leaving it incidental: "What I recommend to the physiognomist as preferable above all is the study of plasters. Nothing is more suitable for observation than a molded figure. . . . These attempts will bring the physiognomist back to the real, to the immutable truths of physiognomy, that is to say, to the study of the solid parts, which will always be the great goal of all his research. One who neglects this basis of our science to focus solely on the movement of muscles resembles those theologians who draw from the Gospel some moral precepts, without recognizing Jesus Christ in them."[9] Through its solidity and its autonomy, a taxonomy of characters constitutes a framework that excludes the study of the signs of passions.

Curiously, this exclusion would be perceived by Lavater's successors as a gap to be filled. It was not until the start of the nineteenth century that we witness a first encounter between physiology and pathognomy. Until then, these two kinds of research had followed separate paths. On one side, anatomy and physiology had pushed as far as possible the examination of muscular structure and movement, in such a minute functional analysis that Winslow was convinced that it might be made to relate to the passions: "One might write an entire Treatise on the almost innumerable combinations of different movements of all these muscles according to the different passions felt by a man."[10] But Winslow never wrote this treatise, perhaps because he was aware enough to see that the project was unrealizable. How could one adjust a finite play of passions to an ensemble of muscular actions that were almost innumerable? On the other side, Lavater had offered a pathognomy, but it remained foreign to muscular physiology. His descriptions stuck to the modifications of features and essentially aimed at character via the fugacity of physiognomic movements. Moreau was one of the first to make a bridge between muscular actions and the expression of passions. He was convinced that it sufficed to animate anatomy in order to explain the nature of passions. Moreau counted fifty-three muscles whose particular movements combined in all manners to produce the play of passions. They were at the disposal of the human soul for the purpose of expressing its different emotions. From the start, Moreau took up Lavater's project and thought he had completed it: "What is proper to us in this supplement is the physiological order and viewpoint in which we have placed . . . the different states of the physiognomy in movement, and all that might be regarded in the exercise of life as the characters of passions." In truth, Moreau was only giving an anatomical and physiological foundation to Lavater's system: "These *observations* are moreover a necessary follow-up to our *Histoire anatomique du visage*, and an indispensable complement to Lavater's research on *pathognomony* [*sic*], which is full of interest, no doubt, and especially attractive for the happy choice of, and ingenious commentary on, a great number of drawings, but by no means indicates precisely (and independently of any influence of the habitual character of the physiognomy) the organic changes, movements, and alterations that constitute the principal trait of each passion."

In the first half of the nineteenth century, the educational role of Lavater's doctrine should not be surprising. The assimilation of the muscular system to an ensemble of apparatuses designated an epistemological threshold. As long as muscular actions were identified with the play of these apparatuses, the observer was confronted with a bundle of simultaneous contractions that were so complex they escaped examination. It was impossible to deduce the expressions from them, which moreover were subject to rapid action and indescribable nuances. To each expression corresponded an ensemble of movements that physiological analysis was powerless to disentangle. Moreau and his contemporaries found themselves in the same situation as their predecessors, but at least they were aware of the difficulties: "These movements, these external changes that reveal with so much eloquence the emotions of the soul, are evidently organic events, physiological facts, for which the various circumstances are, in truth, quite difficult to analyze with precision."[11] Certainly the actions of the fronto-orbicular had been described, and nobody doubted that this apparatus served to express various states of the soul. But one had to be content with indicating, in crude terms, the principal effects of these actions on the face in movement. If it was a matter of all those moderate passions, those agreeable sentiments like joy or serenity, then the face enlarged as if dilated by the cooperation among muscles that constituted the fronto-orbicular: "These *effects* are well expressed in several paintings of the greatest masters of various schools of art; in Raphael's *Saint Cecilia*, in his *Holy Virgin* witnessing the infant Jesus asleep, in the woman present at the death of

Saphire, by Poussin; in the woman placed in the corner of the painting of Darius's Family by Le Brun, etc." If it was a matter of all those concentrated passions in which the fronto-orbicular also participates, then they were depicted by the simultaneous contractions of its various portions: "All this eloquent participation by the frontal muscles in the expression of fright is well rendered on the face of one of the actors in the admirable painting of the enslavement of the Sabines by Poussin. The figure of the satrap, in the battle of Alexander against Darius, also shows well the active part that the muscles of the forehead play in the expression of great fear."[12] Moreau's descriptions remained in the lineage of Lavater's: an incorporeal reality expressed through movements was lodged, without residue, in artistic representation. But the expressive portrait itself remained inexplicable. There are as many passions as individual expressions, and each one of them constitutes its own genre. The transparency of expressive portraits is only the counterpart of a failure. Physiognomy in movement was an opaque structure against which physiological analysis had just foundered.

But not totally: Moreau was quickly engaged in the sole avenue that would provide a physiological basis for expressions. Because exterior manifestations of passions allow common traits to appear, they can be classified. And so physiologists rediscovered the semiology of the soul's emotions. Since antiquity, medical philosophy had divided them into passions that were expansive (warm, gay, or sthenic) or concentrated (cold, sad, or asthenic). When vital energy was augmented, it gave the face a smiling air; when it was diminished, it imprinted an air of despondency on the face. Hence this was a reactivation of the ancient typology of the passions, not without some minor modifications. Sarlandière evoked the class of emotions that express the intellectual faculties (they are midway between the preceding ones) (Plate 3) and Moreau, the class of convulsive passions. Essentially, medical philosophy offered a classification of passions in which muscular actions would come to be enclosed. Far from offering a rule for decoding the correlations between muscular contractions and the play of physiognomy, it allowed them only to be subsumed under its categories. In the expansive passions, movement was centrifugal: muscular actions raise the upper part of the face. When the frontal muscle takes its point of fixed insertion from the epicranial aponeurosis, the skin of the forehead and the eyebrows are raised. Thereby, the face's blooming translates a feeling of happiness. In the oppressive passions, the movement is centripetal: muscular

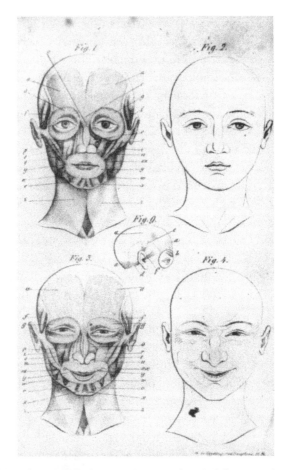

PLATE 3.   Top: drawings depicting seriousness (gravity); bottom: drawings depicting light and gay emotions (joy). From Jean-Baptiste Sarlandière's *Physiologie de l'action musculaire appliquée aux arts d'imitation* (Paris, 1830). Reproduced by permission of Bibliothèque de l'académie nationale de médecine/Université René Descartes, Paris.

actions lower the upper part of the face. The frontal, which contracts from top to bottom, gives the face a somber and melancholic air. The pyramidal, which is attached to the bone of the nose, acts powerfully in this lowering of the frontal. There results the sentiment of great pain in the soul, the expression of oppressive passions. Other muscular movements are governed by the same principle. When the parts of the face move away from the median line, an expansion appears. In laughing, the dominant trait results from the horizontal contraction of the zygomatic muscles. But when the parts of the face move closer to the median line, concentration appears. In the violent passions, the dominant trait results from the horizontal contraction of muscles of the superciliary part.

The paradox lies in the very principle of characterizing passions. In its singular form, each passion is the result of a combination of muscular movements that cannot be examined. Physiological analysis renounces its rights: each passion is characterized only through its physiognomic effects. Moreau evokes a gallery of expressive portraits. But when it comes to classifying passions, the division is based on a principle that arises from organic spatiality. The actions of the parts of the face and the underlying muscles are called upon to serve as points of coherence for the ungraspable diversity of the play of physiognomy. If the latter may be organized in such a way as to form general modes of facial expression, this is because they present signs that are all the more eloquent in that they are matters of more or less surface. More surface: the expansion of parts was the major symptom and most certain sign by which one might recognize that series of passions that were called "expansive"; less surface: the tightening of parts was the sign closest to the essential thing by which one could identify that series of passions that were called "oppressive." There was only one way, then, to apprehend the characters of passions: to consider them in a general way, relative to the nature of signs that physiological analysis untangled in their expression. Moreau thus returned to the Cartesian distinction between instinctive and intentional movements. But in squaring it with the division that Bichat had just made between organic life and relational life, he modified their status significantly. On the one hand, involuntary signs that depend on organic life are symptoms. But instead of testifying to passions, they betray them in vehement, useless, and uncontrolled actions. On the other, voluntary signs define a language arising from the life of relation: to live is to move and express oneself. There is a reflected use of signs in order to represent an interior state. An assumed facial expression (*mimique*) is the mark of and substitute for thought, or rather, for a thought that is inscribed within life. By this relation of a language without speech to the life of relation, facial expression's artifice is grounded in nature: "Its play, its movements alone constitute the detailed and voluntary gesture of the face. . . . This apparatus is the organ of a particular locomotion, more delicate, less extensive, by which man . . . asks all that surrounds him to serve his will, to listen to his thought, to respond to his affections."[13] Although Moreau was pointing to the interplay of physiognomy and muscular actions as an *expressive language*, he could decipher neither the text nor the mechanism.

Yet Duchenne thought he had found in Moreau the first descriptions of a language and a physiology of the passions: "He addressed the particular and detailed examination of the action and effect of each muscle on facial expression. He touched on physiological questions, which according to this author (and this is also my opinion), had never been tackled in the best treatises of anatomy and physiology published before him."[14] In reality, physiological analysis merely allowed one to classify a passion that gave itself an ensemble of signs as the characteristic symptoms of its essence. Physiological analysis had just situated itself, by way of a complementary examination, inside a semiology whose signs referred to the essence of passions. These latter did not draw their origin or form from physiological movements. The causal relation did not move from muscular actions to the play of physiognomy, but from passions to the linear and fixed language of physiognomy. In other words, the repeated contractions provoked by natural penchants, that is to say the dominant passions, ended up giving the face of a subject the imprint that *announced* him or her. "The *effect* becomes here a symptom; it reveals and manifests its cause through the signs, of which sustained experimentation and numerous and careful observations make recognized and appreciated all the degrees of value and of expression."[15] Passion is deposited in the organism and inserts its signs there. It is itself inscribed at the level of the face by means of muscular contractions. But its essential structure remains a precondition. The muscular system of the face is provided with references to this structure: it *signals* it but does not order it. The characters of the passions drawn from muscular actions are the expression of character. The interest aroused by Lavater's aphorisms relates to the fact that they procure the *fixed lines* on the basis of which it was thought one could decode the play of physiognomy. Lavater identified the *essential*. For example, if the mouth is tightened and rises a little at the extremities, it announces a fund of affectation, pretension, and vanity. This aphorism, like so many others, seemed to reenter the domain of the muscular physiological: the repeated contraction of zygomatic muscles brings this trait to the face. Lavater decoded character on the basis of signs; Moreau saw in them the traces imprinted by repeated contractions of muscles under the rule of character. Apparently, Moreau was explaining immobile physiognomy by physiognomy in movement, but in fact he passed from facial expression, which referred to the basis in character, to the play of physiognomy, and inversely. There was perfect circularity.

Let us recall briefly the triple-barrel lock that made physiological analysis dependent on physiognomical doctrine. First, fibrillary theory included the perception of synergetic movements. Each expression requires the cooperation of an ensemble of muscles. With the *principle of cooperation of muscular actions,* there was still the question of the action of expressive muscles. Then, each folding of the skin presupposes the contraction of the underlying muscles. With the *principle of bi-univocal correspondence between folds and contractions,* the face was perceived as a simple surface on which muscular actions were inscribed. Finally, the analysis of the pathognomic signs of passions oscillated between an iconographic description and a typological examination. With this *principle of identification of passions,* the signs were recognized without being understood. It was by rejecting these principles that Duchenne effected a triple unlocking. He thus disengaged the study of the emotions from the framework in which it had been caught: a metaphysics of passions. To grasp the significance and bearing of his method—indistinguishably, electro-physiological and icono-photographic—we must truly see the difficulties he was confronting in 1856: no longer fibrillary theory, since it had been refuted in 1850, but its three corollaries. This was because their validity appeared still well founded—if not in theory, then at least in perception. It suffices to apprehend a harmonious expression, after the excitation of a single muscle, to understand that it springs from an optical illusion. Hence the rejection of the principle of cooperation among muscular contractions. It suffices to see, after the excitation of a single muscle, a series of folds where muscles had not been stimulated to challenge the mechanism imagined by Camper. Finally, from the moment when each expression could be related to the muscle that provoked it, the uncertainties disappear that were linked to the sole perception of expressive portraits or to the general allure of a passion. Pinning an expression down to the passion to which it corresponds would require a demonstration that carries a degree of certainty. Three blind spots had compromised study of the passions, and the forthcoming transformation would derive in an interconnected way from the position of the observer, the choice of subjects of study, and the mode of decoding passions. A new field of visibility emerged: the study of the physiognomical mechanisms presupposes that the effects of the localized electrization could be seized by a medium.

Taking Pictures

We have to start with photography, which in this period was traversed with the positivity of a body of knowledge. Technical improvements were responding to a series of problems. In 1839, François Arago had revealed the daguerreotype method, and the procedure was widely used until 1850. The use of accelerating substances allowed an improvement in the sensitivity of the plates, but posing times remained long. The year 1851 was marked by the discovery of collodion, which would revolutionize photographic practice. This innovation was at the intersection of two types of research, one concerning the negative and the other the support material and accelerating substances. At issue was the ability to obtain reproducible images and to reduce the exposure time during shooting. Louis Blanquart-Evrard perfected a procedure that had been proposed in 1840 by William Henry Fox Talbot for realizing negatives on paper, called calotypes. Based on the notion of a latent image, the calotype enabled a paper negative to be obtained, which offered the advantage of conserving sensitivity over time so that the image could be developed later. This discovery allowed the two operations to be separated: the shooting of the image and the development of the negative. C. F. A. Niepce de Saint-Victor replaced the negative paper with glass, which enhanced the quality of the proofs. He used albumin in order to retain the chemicals on its surface. Thus the quality of the prints could rival the daguerreotype in clearness and precision. But albumin-treated glass required an increased posing time, and thus photographic studios retained the daguerreotype process, resorting to hand engraving in order to publish the portraits. The invention of collodion was decisive. In 1831, the Englishman Frederick Scott Archer published his *Manual of the Collodion Photographic Process*. Collodion, or cotton powder, dissolved in ether alcohol, was a sticky substance that adhered to the glass and absorbed by soaking up the silver salts. This material, as long as it was humid, took an impression from contact with light rays, and so rapidly that movements could be instantly recorded. Auguste Adophe Bertsch improved this procedure in 1852. The time it took to make a portrait in good diffuse light was only between two and four seconds, and from five to twenty seconds in weak diffuse light. Associated with positive printing on albumin-treated silvered paper, collodion allowed the production in series of commercial images for the public at large.

Duchenne had undertaken his electro-physiological analyses at a time when developments in technology allowed photos to be taken without excessive posing times—hence he could extend his experiments by making iconographic studies. But recording the expressions of subjects through photography was not without difficulties. Adrien Tournachon, the brother of Félix Tournachon (whose pseudonym was Nadar), had taught Duchenne the theory and practice of photography, even the ways of using lighting. Adrien was already well known for having placed his models in full light so as to reduce posing time. This conferred upon portraits an incomparable truth. His studies of expressions of the mime Jean-Charles Deburau in *Pierrot* (1854) were very successful at the Universal Exposition of 1855. Ernest Lacan reviewed it with effusive praise: "M. Tournachon and Company have gathered, in a series of prints showing the mime Deburau (in the traditional costume that his father made famous), a series of diverse expressions that they have rendered with great skill. Here, Pierrot is under the influence of a wave of hilarity; his mouth stretches to his ears. . . . There, he is under the impression of fear, his great white face elongates. . . . Farther on, Pierrot listens and his whole person expresses attention; then it is surprise, then pain, or drunkenness or greed. Each of these prints is a study of an admirably rendered expression, which gives as much honor to the talents of the photographers who have executed it as to the spiritual artist who served them as model"[16] (Plate 4). On the basis of this collaboration between Adrien and Duchenne, some historians have attributed the paternity of a great number of prints to the former—notably, the photographs in which the old cobbler and the experimenter both figure. The reasons they invoke are stylistic, but also the fact that Duchenne's personal *Album* contains some photographs signed "Nadarjne," which was Adrien Tournachon's pseudonym. In fact, a recent inquiry into the photographs conserved in the National Archives would tend to confirm Adrien's preponderant role. Five boxes containing the photographic images that accompanied Duchenne's paper for the Volta Prize have recently been found: "The boxes of photographic plates are

PLATE 4. Gaspard-Félix Tournachon (Félix Nadar) (1820–1920), "Pierrot souf-
frant" (1854–1855). Musée d'Orsay, Paris. Photograph copyright RMN. All rights
reserved. Reproduced by permission.

printed with the initials 'A.T.' referring to Tournachon, followed by '*Nadar
Jne.*' The photographic stamp reproducing in manuscript characters *Nadar
Jne* [*jeune* = younger] is moreover placed on 13 of 32 prints. . . . On some
plates, Duchenne takes care to cross out the printed initials and pseudonym
of Adrien Tournachon and replace them with his handwritten name."[17]

Hence there is a paternity problem. Should we attribute the prints
to Adrien Tournachon or to Duchenne? To answer this question, we must

apply the criteria of the period with respect to artistic property. What did Nadar say in the lawsuit pitting him against his brother Adrien? "I propose to turn your coachman and your concierge—I am speaking seriously—into two more photographic operators after only one lesson. Photographic theory can be learned in an hour; the primary notions of practice in a day. . . . What cannot be learned, I will tell you . . . is the psychological side of photography."[18] Duchenne was saying the same thing as Nadar, since he distinguished the camera operator from the author who directed the operations of shooting: "Art does not rely only on technical skills. For my research, it was necessary to know how to put each expressive line into relief by a skillful play of light. This skill was beyond the most dexterous artist; he did not understand the physiological facts I was trying to demonstrate." While Duchenne makes it clear that he has himself photographed most of the figures, or "presided over their execution," he also recognized his debt to his friend, the brother of Nadar: "M. Adrien Tournachon, a photographer whose ability is known to everyone, has been kind enough to lend me the sum of his talent to execute some of the negatives for this scientific section."[19] So we have the following alternative: either Duchenne presided over the execution of the negatives and the able Adrien was merely the executor, or else Adrien is the author of the photos and we must exclude the possibility that Duchenne presided over them. At least two reasons permit us to dismiss Adrien as responsible for them. From the moment he appropriated the pseudonym of his brother, it is legitimate to think that he might also sign photographs of which he was not the author, and the lawsuit won by his brother in 1857 lets us suppose this was so. The second reason is determining. No doubt Adrien did give good advice to Duchenne; concern about lighting was still commonplace at the time. Francis Wey had already recommended that photographers take inspiration from the paintings of Rembrandt, Rubens, and Ribera. Without doubt, Adrien had photographed some electro-physiological experiments, but he could not have taken the least initiative with regard to the subject of this photography: the disposition of the model, the framing of the experiment, the details to be highlighted, and the moment of releasing the shutter. This moment was doubly crucial: first, it carried all kinds of signification (anatomical, physiological, and physiognomic), and second, it established the full responsibility of the doctor of Boulogne.

Curiously, historians of photography have largely contributed to occult Duchenne's preponderant role—to the point of eclipsing his authorship in an enterprise that was wholly his own. In centering their interest on the Nadar brothers, they have ended up blurring an issue that could have been quickly settled. Here lies another paternity issue: André Jammes once launched the myth of the supremacy and omnipresence of the great Nadar: he claimed that Adrien initiated Duchenne and that Nadar advised Adrien. Hence the Boulogne doctor was placed in Nadar's shadow: "Even when Nadar remained in the background, his influence sufficed to guide the executor and to give spirit to images that without him would have remained banal. Dr. Duchenne's album perhaps benefited from the presence, physical or occult, of the great Nadar. When the physiological images take on a truly tragic dimension, it is tempting to discern in them the hand of a master."[20] In the images of *Mécanisme de la physionomie humaine*, the Nadar brothers' savoir faire is meant already to show through. Recently, Françoise Heilbrun stressed the ascendancy of the Nadar brothers over Duchenne. She has even discovered a stylistic affinity between the series of photos by Duchenne and those in the *Pierrot*. The expert intervention of the Nadar brothers on Duchenne's elderly model is said to have preceded and even prepared the way for the pictures taken of the mime Deburau. "Is Nadar the source of Adrien's work? We have this clear feeling without being able to provide decisive proof. In any case, Duchenne's series served as field of experimentation for staging the series of studies of *Pierrot*. . . . One notes the almost baroque virtuosity with which the space is treated, into which the artist draws us by a play of hands as graciously put together as a ballet. We will find this hectic vivacity of composition again in the series *Études d'expression de Charles Deburau en Pierrot*."[21] Sylvie Aubenas, in turn, follows Heilbrun when she analyzes "The Interview of Chevreul" (1886): "In this last collaboration between art and science, one finds an echo of the first photographic attempt by Nadar, when he advised Adrien, his brother, on the pictures taken for Duchenne of Boulogne in 1854."[22]

This version is historically false. We have to return briefly to its methodological presuppositions: an iconographic analysis lacking rigor and a fantastic dating. First, there is a confused perception of the images: the comparison between an experimental gesture and a pantomime arises from pure fiction. To establish a relationship between the *Pierrot* series and the images of *Mécanisme de la physionomie humaine*, it would have been necessary to identify some indices, but a quick examination shows instead the immense distance that separates them. *Pierrot* is a role, a mask, and a mime who "speaks." Duchenne's models are experimental subjects, without makeup, whose expressions result from manipulation. *Pierrot* is dolichocephalic and his facial angle close to a right angle. His forehead is high and domed; his nose long and straight; his mouth big and his dentition perfect. If the cobbler is close to a type it is rather the clown, the actor who plays the role of the ridiculous old man. Then, Pierrot's makeup eliminates all effects of coloration, inasmuch as its function is that of the masks worn by ancient actors. But it also dissimulates all facial defects. Charles Hacks has stressed this essential point: if the powder is badly applied, "horrible wrinkles, ducks feet notably, and the naso-labial furrow will be seen underneath the white and will be irremediable."[23] Duchenne was interested in natural faces, and all that would disappear under the *Pierrot* makeup must appear in his images. Similarly, anything that is retouched, like the eyebrows traced in charcoal and the shape of the mouth, should be manifest without dressing: the hairy eyebrows and the various movements of the corners of the mouth. The *Pierrot* costume is ample but tight at the neck, so he can sweat at ease, amplify his movements, and make his gestures stand out. The white shirt of the cobbler is wide open at his chest, to show the skin of his neck. Duchenne has made a direct (though veiled, out of regard for the Nadar brothers) allusion to what distinguishes their model from his own: "I preferred not to choose my models as artists often do, when they select subjects with facial features suiting a particular expression. . . . Moreover, not wanting to combine gestures with the facial expressions of my subjects, I have given all my models in this scientific section the same pose. In spite of these restrictions . . . the artificial expressions that I photographed remain grippingly true."[24] In short, the two series of photos have nothing in common. Duchenne sacrificed finish for exactitude of expression and did not judge it useful to proceed with retouches. Photographic imperfections did not prevent recording the essential thing: the clearness of the expressive lines. Duchenne pursued a scientific goal that was the opposite of the aesthetic preoccupations of the brothers Nadar.

As for the error in dating the photos of the old cobbler, it results from an a priori historical discrimination. To place Adrien's collaboration with Duchenne under Nadar's impalpable supervision, it would be necessary for the photos to have been taken at the time of the collaboration between the Nadar brothers—hence in 1854, the moment when Félix comes to the aid of his brother Adrien in bankruptcy and invites the mime Deburau to come and pose in his brother's studio. It was not before 1854, since that is the apprenticeship period of the Nadars, nor after, given the conflict that breaks out between them and leads to the lawsuit of 1857. Given those parameters, historians have believed that they found in Duchenne's prose the evidence that would allow them to assume that 1854 was when the cobbler was photographed. In the foreword to the "Scientific Section," he writes: "From 1852, convinced of the possibility of popularizing or even of publishing this research without the aid of photography, I approached some talented and artistic photographers. These first trials were not, and could not be, successful."[25] On the following page he is specific about the period in which his pictures were taken: "from 1852 to 1856." That Duchenne has given this specific information—or rather, this approximation—does not confer on it any privilege of authenticity. It is possible that he turned to artists around 1852, but these artists were not photographers. In "The Purpose of My Research," Duchenne lets slip a piece of information that contradicts the preceding one. The artists to whom he appealed were *plastic artists*. "Skillful artists have tried in vain to represent the faces of my subjects; for the contractions provoked by the electrical current are of too short a duration for an exact reproduction of the expressive lines that develop on the face to be drawn or painted."[26] There is more: in 1855, Duchenne had still not elucidated the various movements of the eyebrow, for which we recall the reason. The superciliary, because it was still perceived as depending on the orbicular of the eyelids, could not be identified as the muscle of pain. Similarly, at this time, Duchenne believed that laughing depended only on the contractions of the grand zygomatics. The doctor of Boulogne was very well aware of the limits of his earlier work: "What share should we give to each muscle of the face regarding the influence it exerts on the play of physiognomy? I was far from having decided about these complex and difficult questions that I had barely touched upon."[27] Duchenne perhaps called upon sketch artists to represent the simplest expressions, like crying and attention. But the frank laugh—and especially pain and all the expressions in which the superciliary participates, like sadness, fright, and terror? It is clear that all could not even have occurred to him.

The chronology cannot be modified: in choosing between the period 1852–1856 and the year 1856, the year 1856 must be chosen for *all* the photographs. It has scarcely been noticed that Duchenne linked the history of the images of physiognomy in movement to that of his pathological images. In 1862, he published his *Album de photographies pathologiques*, in complement with the second edition of *L'électrisation localisée*. In the foreword to the *Album*, Duchenne writes: "Photography alone can show Nature as she is observed in these kinds of pathological manifestations. Starting in 1852, I had already had the idea of representing, with the aid of this marvelous procedure, the specific action of muscles excited individually by localized electrization, and the functional troubles that one observes consecutive to muscular afflictions, either at rest or during movements."[28] Once more Duchenne blurs his tracks. In effect, if his interest in photography dates from 1852 and if some of his shots were taken prior to the year 1856, one ought to find some prints in the first edition of *L'électrisation localisée* (1855), and these prints ought to be of the same kind as found in the second edition (1861): not just clinical descriptions of some of the photographed cases, but the precision that would mean the drawings illustrating these cases had been done from the photographs. The results of a comparative analysis between the sixteen pathological shots and the 1855 edition are eloquent. Figures 1, 2, 3, 4, 12, 13, 15, and 16, from pictures stretching from 1856 to 1861, are automatically excluded. Figures 8 and 9 correspond to two cases for which he did not give a clinical history. Figures 5, 6, 7, 11, and 14, undated, concern a series of clinical cases that are not described in the 1855 edition. Only figure 10 remains (Lucien Gaulard, age 32, porter). The clinical description of this case is indeed found in the 1855 edition, but *without illustration* (Observation 189, pp. 824–826). On the other hand, in the 1861 edition, the clinical story of this case is accompanied by a drawing done after the photograph in the *Album* (Observation 100, fig. 65, pp. 444–455). Thus the Gaulard case, described before 1855, was photographed later. As a final precision that complements the preceding one: in chapter 4 of the 1855 edition, titled "Electro-physiology of the Facial Muscles" (pp. 373–392), Duchenne does not make the slightest allusion to photography. From this examination, one may conclude that he did not practice photography before 1856, and that the idea of doing so had not even touched his thinking.

But what was the purpose of establishing by factual proofs that 1856 is indeed the year of the first camera pictures? Of situating the event, at that moment, in the direct line of a central problem that Duchenne had just resolved? This was not by chance, for he had turned to photography after a fundamental discovery. As long as he was ignorant of the expressive functions of various motor muscles of the eyebrow, the *motifs* for photography did not exist. It would have to wait until 1856, as we have seen, for the conditions of possibility, both technical and epistemological, to be conjoined that would allow the movements of this facial part to be at last disentangled. Once again, we must recall the determining role played by his first experimental subject, the old cobbler affected by facial anesthesia. With him, Duchenne solved (or rather, definitively elucidated) the ensemble of questions relating to the anatomo-physiology of the fronto-orbitary region. Despite his previous work, Duchenne truly had the feeling of *inaugurating* his research on physiognomic movements with the cobbler: "On this subject I did a great number of icono-photographic experiments, when I *began* my research on the mechanism of human facial expression."[29] This is a fact. The motifs of the photography are not faces, but biological events: muscular contractions, physiological actions, and expressive movements. The latter are not pre-given but discovered in the old cobbler in 1856. Until then, Duchenne considered that the superciliary was only a *dependent* of the palpebral orbicular. Thus the decoupling of the superciliary unblocked the epistemological jam. The unpublished paper of 1857, "Anatomical and Experimental Research on the Motor Muscles of the Superciliary," offers direct testimony on this. From the moment Duchenne separates all the motor muscles of the superciliary, he is able to attribute to each of them an expressive function: "In summary, the electro-physiological experiment demonstrates that there exists a point of separation between the superciliary (muscles of suffering), the frontal (muscle of attention), and the superior palpebral orbicular (muscle of reflection)."[30] From the study of the superciliary also flowed the rules for analyzing physiognomic movements.

Optical Illusions

An initial problem posed to psychology concerned the relation between an expressive movement and the muscular contraction that corresponded to it. From the instant that electrization of a single muscle provoked a harmonious expression, one could no longer fall back on a synergetic movement. If the new certainty was on the side of *localized* electrization, then one had to challenge visual perception itself. Not only had the principle of association of muscular contractions rested on a false theory, but also it derived from an optical illusion. The action of the superciliary always produced an expression that seemed to mobilize different muscles. The superciliary took the form that characterized the expression of suffering. Moreover, the movement of the mouth and the naso-labial line harmonized with it to depict this painful state of the soul. It sufficed to experiment on recently deceased subjects to produce a harmonious expression by the excitation of the superciliaries alone. No reflex contraction: the illusion is complete. Only experimenting on a cadaver could raise the suspicion that an optical illusion was in play, and the idea of how to dissipate it: "It came into my head to leave uncovered only the lower part of the face while I excited the muscles of pain. Imagine my surprise in seeing that this lower part of the face did not show the least modification in dead subjects."[31] But Duchenne inverts the order of surprises. What astonished him was not just that the uncovered part of the face presented no modification, but rather that the face of a dead person offered the aspect of a modification of the whole upon the excitation of the superciliaries alone. This point is essential: an *expressive* movement that springs from an illusion can no longer be confused with a *synergetic* movement. There is a new distinction between the real movement and the apparent movement, between the modifications of a superciliary provoked by the contraction of its muscle and the perception of an ensemble movement. The method of exposition reverses the usual approach: first describe the physiognomic movement such as it appears, and then the mechanism that makes the expression seem apparent. Very quickly Duchenne would discover the ensemble of completely expressive muscles: the frontal (muscle of attention), the superior palpebral orbicular (muscle of reflection), and the pyramidal (muscle of aggression).

But other expressions depend on the combined action of several muscles. Here their law of composition had to be found. To do so, Duchenne first identified the incompletely expressive muscles. He described expression as it appears: a modification of the whole face. Then, he showed that this expression arises from artifice (and no longer from an optical illusion): "These muscles are highly expressive. . . . Nevertheless one then notices that the expression is not natural, that it is artificial, and ultimately that it lacks something."[32] Solely by the contraction of the major zygomatics an artificial gaiety appears. For this borrowed expression to become a natural expression, it must be associated with the action of the inferior palpebral orbiculars: laughing is then sincere and spontaneous. But sometimes the expression linked to the contraction of a muscle may be undecided. For example, the transversal of the nose expresses lubricity just as much as ill humor. In order for an expression to end up as pronounced, one must find the muscles that act in association with it. By its combination with the muscle of joy, the contraction of the transversal comes down in favor of lasciviousness. By its combination with the contractions of the sphincter of the eyelids and the triangular of the lips, there appears an expression of contempt. Finally, sometimes the same muscle changes its status according to the combinations in which it is participating. For example, the contraction of the lower half of the palpebral orbicular gives an expression of benevolence; here, it is a completely expressive muscle. However, when associated with the zygomatic major, its contraction contributes to the expression of frank laughter; in this case, it is a complementarily expressive muscle. Essentially then, expressions may be produced either by completely expressive muscles or by the incompletely expressive muscles in combination with the complementarily expressive muscles. These belong to the category of primordial expressions.

Finally, it is by an association of primordial expressions that complex expressions may appear. Duchenne here resolved two problems that Denis Diderot had raised. The first was posed by Carle Van Loo's painting *La chaste Susanne*. The critic found that the old men lacked heat; more desire was necessary and more élan in their expressions. The solution lay in the production of a corroborating complex expression. This was obtained by the addition of traits specific to various passions. Attention is produced by the contraction of the frontal and joy by the combined action of the major zygomatic and the inferior palpebral orbicular. By associating these primordial expressions, physiognomy could announce that the soul was under the lively impression of an unexpected happiness. It suffices to add the sign of lasciviousness to obtain a new expression. "The sensual traits belonging to this last emotion will show the special character of attention brought on by something that excites lubricity. It will portray perfectly, for example, the faces of the lewd old men spying on the chaste Susannah."[33] The second problem concerns the production of a mixed expression.

Diderot wished that painters could manage to represent two actions of the face: "It is neither counter to the truth nor counter to the interest to recall the instant that is no longer, or to announce the instant that is going to follow, . . . as on a face where pain reigned and where joy is going to be born, where I will find the present passion confounded with the vestiges of the passion that is passing away."[34] The solution resides in the association of two primordial but contrary expressions, which put into play antagonistic muscles. This is a "complex discordant expression," which does not signify that the observer is confronted with grimacing forms. The whole difficulty consists precisely in making a division between the expressive and the inexpressive. The contrast between a grimace and a harmonious expression amounts to almost nothing: the degree of contraction of antagonistic muscles is what makes the difference. Is the excitation of the muscles of joy and pain energetic? Then the expression tilts toward the inexpressive, and it is a grimace, which is scarcely different from a muscular spasm. Is the contraction of the same muscles moderate? The lines that express joy are harmoniously associated with those of pain. Duchenne could recognize the image of a melancholic smile: a ray of joy that yet cannot dissipate the traces of a recent pain.

The Emergence of Variables

A second series of questions is posed when psychology matches the displacements of the skin to the action of muscles. How to explain the variations in what is modeled that result from the excitation of the same muscle in different subjects? One would have to take into account the appearance of cutaneous furrows that may be more or less numerous, and more or less distant from each other. As soon as an expression is no longer related to a synergetic movement, one must revise the description of the way in which folds are formed. These modifications in the face must result from a mechanism different from what Camper had imagined. Not only did the principle of binary correspondence between the folds of the skin and the contraction of underlying muscles rely upon a theory that was false, but also it arose from a vicious circle of reasoning. Camper intuited the direction of the muscular fibers from that of the lines and vice versa. By ascribing to physiognomy the actions of individual muscles and by comparing the skin to a tissue, Duchenne broke this circle. What does one see when one applies vertical or oblique traction from bottom to top on a point of the surface of a curtain whose fabric is supple? Depending on the texture of the cloth, pleats form in greater or lesser number in various directions and on different points of this surface: "Similarly, on the surface of the facial skin, one often sees certain lines or furrows forming in various directions and placed more or less distant from each other, under the influence of a simple traction exercised on a surface point."[35]

Duchenne did an analysis of the various factors capable of modifying facial features: in particular, the age of the subjects and the effect of external modifiers on the skin. In the same experimental conditions, the differences that affect the texture of the cutaneous envelope can account for physiognomic variations. Comparative study of the effects of the same muscular excitation in different subjects proved decisive for obviating all difficulties. Diderot was wrong to see, among both infants and the elderly, an unformed mass tending to development or to be reduced to nothing. Age is a differentiating factor. This is why Duchenne chose as experimental subjects two children, a young man, a young woman, and an old man. Among extrinsic factors, he paid attention to climate. He chose a woman aged forty, with skin tanned by the sun. Recording the modifications observable among his various subjects led to a distinction between principal and secondary lines. The principal lines appear in a repetitive form, designating what was essential, the phenomena constant in and common to all experimental subjects. Secondary lines refer to constant modifications that are observable among only aged subjects or those whose skin is suntanned. "The traits belonging to a particular expressive movement are composed of *fundamental* lines, which make up their pathognomic signature, and lines which I call *secondary*. Secondary lines may be missing in certain conditions; and if they appear, it is only as satellites of the fundamental lines, adding to their significance, altering the degree of the emotion, hinting at the age of the subject, and so forth."[36]

With the little girl, the contraction of the frontal provoked the swelling of the head of the eyebrow marked by a gentle curve: this is a fundamental line. In the old man, or the woman whose skin is sunburnt, secondary lines can be seen. To the swelling of the head of the eyebrow is added frontal pleats, curvilinear and concentric with the arc of the eyebrow, which might be gathered and continued on the median line. The appearance of lines more or less distant from the electrically stimulated muscle arises from the same explanatory principle. What do we see after the excitation of just the muscle of joy in an aged subject? The corners of the lips rise, the naso-labial furrow deepens, and its curve becomes sinuous; finally, radiating lines appear in the perimeter of the external angle of the eye. Moreau, who applied Camper's criterion, thought he was recording the synergistic action of three muscles: the commissure is put into movement by the zygomatic major, the naso-labial furrows deepen under the influence of the elevator specific to the upper lip and the nose, the lines radiating from the external angle of the eye are due to the action of the sphincter of the eyelids. In reality, all these modifications result from the zygomatic major. One must relate this muscle to the elevation of the labial commissures, the deepening of the naso-labial furrow, and the radiating wrinkles at the corners of the eyes. Crow's-feet, which appear only in the adult, are all the more marked in subjects who are aged or sunburnt.

Pedagogy of the Gaze

A third and final problem was posed if psychology wanted to identify each emotion by its expression and to fix a recognizable image of each expression. From the moment when the expressive movements are founded on knowledge of the muscular mechanism, the old typology of passions and examples drawn from Beaux Arts lose their credibility. Not only was the principle of identifying the passions based on a false theory, but it had been based on a haphazard phenomenology. Sometimes certain typical movements were used to designate transformable structures, from which classifications were derived according to their vague content (expansive, oppressive, convulsive passions), and sometimes a painting was offered as model for a passion. But in either case, the criterion for identifying the expressive kernel was founded on a perception whose over-determination was considerable. Perception put into play the observer's intellectual capacities, as well as taste and imagination. The identification of passions constituted an open and fertile, but fundamentally uncertain, domain. In 1857, Duchenne believed that the presentation of photos alone would suffice to carry conviction. The observer would be confronted with troubling images of truth that were easily identifiable: "Photographed figures that represent, as in nature, the expressive trait assigned to the muscles that interpret the emotions, teach a thousand times more than extensive written descriptions"[37]—hence the idea of publishing his general introduction and the album of photographs with some explanatory notes. But his failure in the competition for the Volta Prize (1857) made him change his mind. In the opinion of the commission, the general introduction and the album were *insignificant* in the proper sense of the term. Duchenne learned his lesson: in 1862, in *Mécanisme de la physionomie humaine*, he added the "Scientific Section," containing explanation of all the facts highlighted in his electro-physiological experiments: If the truth of images is linked to the demonstration of truth, then everyone will be led to form the same judgment. The reader is an observer, too, and deduces the significance of the image from the laboratory notes and follows the demonstrations on the basis of the image. If the photos speak the same language to all, then it is because each person is able to follow the text to which the photos give light. It is simply not true, as Mathias Duval wrote, that before these diverse faces "any judge, artist or physiologist, expert or not, educated or not, immediately gives the verbal translation of an expression that even a child understands without hesitation."[38]

Around the photo, Duchenne organizes a demonstration that follows an unvarying pattern, moving from description of the expression as it is perceived to the muscular action on which it depends. But the photograph also records experimental gestures, since without them there would be no image. The camera operator, the stimulating electrode, and the expressions of the experimental subject: everything is linked and mixed together. The shot presents the experiments and their meanings, on condition that the observer has access to all the elements in order to appreciate the value of the experimental procedures and their results. If so, then there is no difference between the study of the mechanism of facial expressions as knowledge and as teaching. The effort of the experimenter to know, and the act by which the reader learns, are accomplished at the same time. The physiognomic experiment, in its double aspect of manifestation and transmission of a body of knowledge, has a collective subject: the experiment is simultaneously performed by the person who demonstrates the mechanism of expressions and those before whom it is unveiled. One better understands why the Volta Prize Commission in 1857 could not really grasp the significance and import of the photographs. "The representation of these phenomena by photographic pictures is a new and felicitous idea whose results he will make known later. The commission has in fact only ever received these drawings."[39] And so, with the publication of his book in 1862, the album was accompanied by its indispensable complement: the "Scientific Section." Because the photographs are linked to the demonstrations, they preclude in advance all interpretations by the observer (or his puzzlement); he should not lose sight of what Duchenne asks him to see—what he is demonstrating. Darwin, one of the most attentive readers of *Mécanisme de la physionomie humaine,* understood this very well: "When I first looked through Dr. Duchenne's photographs, reading at the same time the text, and thus learning what was intended, I was struck with admiration at the truthfulness of all, with only a few exceptions. Nevertheless, if I had examined them without any explanations, no doubt I should have been as much perplexed, in some cases, as other persons have been."[40]

Fixed by photography, passion has a materiality, a support, and rules for decoding it. One must distinguish the different levels of interpretation. An "interior calm" corresponds to the absence of expressive movement. The equilibration of forces designates an active immobility that bears the mark of modifications shown by the tonic force of the muscles after the play of the usual emotions. The absence of expression is the face in repose, already a character. The portraits of models show this. By comparing muscular actions, a second level of meaning appears. For example, the frontal is a completely expressive muscle (the muscle of attention), but the muscle of the neck is an inexpressive muscle: "I have opened the subject's mouth while stimulating [it], but the result has been a grimace or a deformity analogous to that of burn scars in the cervical or thoracic region." By comparison of combinations of muscular actions, a third level of signification appears: for example, the various degrees of intensity of an emotion (Plate 5). Fear (fig. 60): combines contraction of the muscle of the neck and the frontals. Terror (figs. 61, 62, and 63): contraction of the muscle of the neck, frontals, and lowering of the jaw (Plate 6). Terror combined with pain (figs. 64 and 65): combined contractions of the preceding muscles with that of the superciliary. Finally a brief history allows the reader to apprehend the cause of the passionate movement and the response of the subject who *faces* it. The stories imagined by Duchenne give the image the character of a lived experience: "Figure 60 is the image of fear. . . . We do not see, at least in his face, that this individual is really in danger; we only sense that he dreads

PLATE 5. Stronger electrization of superior orbicular palpebral, with slight lowering of labial commissures: meditation, contention. From Guillaume Duchenne's personal album, fig. 15. Reproduced by permission of École nationale supérieure des beaux-arts, Paris.

something. But looking at figures 61, 62, and 63, doubt is no longer possible; this man is frozen and stupefied by terror; his face shows a dreadful mixture of horror and fear, at the news of a danger that puts his life in peril or inevitable torture . . . but in figures 64 and 65, the horrible pain of torture has been added to the expression of this terrible emotion. This expression must be that of the damned."[41]

In short, only photography makes possible the examination of the face, the examination of indices, and the emergence of the subtlest nuances. Objects of unleashed perception, the shots allow one to study lines and contours—the multiple and infinite modifications of the face. The photographs themselves are objects of study. Duchenne rightly spoke of his "icono-photographic experiments"—hence the enigmatic title he inscribed on the inside of the boxes in which he stuck the prints that accompanied his 1857 paper: *Opséoscopie électrique.* The term (*opsis,* genitive *opeus*) might have several meanings, but given the era in which Duchenne forged this neologism, it probably means "optics." This term designates the photographic apparatuses and applies also to the prints. This title offered the advantage of indicating the object aimed at, the very materiality of the shot. *Opséoscopie électrique* was the examination of photographs of a body subjected to localized electrization. This etymology is moreover confirmed

PLATE 6. Electric contraction of the muscles of the neck, the frontals, with a voluntary lowering of the lower jaw, and lowering of the upper eyelid and a downward glance: expression of fear. From Guillaume Duchenne's personal album, fig. 63. Reproduced by permission of École nationale supérieure des beaux-arts, Paris.

by the caption that accompanies the photography illustrating his book: "Photographic Electro-physiology." But the disadvantage of using these terms is apparent: it leaves aside the *motif* of this kind of photography, expressive physiognomy. This is no doubt the reason why Duchenne preferred the title *Mécanisme de la physionomie humaine.* But now it is the essential function of the image that is occulted. It would not be false to say that Duchenne represents, by means of photography, the effect of muscular contraction (the expression that results from it)—but it would be highly insufficient. If the anatomical study of passions were reducible to electro-physiological analysis, photography would be only the recording of an experiment. But the shot is something quite other. What it fixes is the appearance of a face, as one perceives it after a muscle is stimulated. On this basis the enterprise of observation, description, and comparison of images could begin. Hence the role, once again decisive, of instrumental mediation—in this case, optics. In Duchenne's studies, techniques, methods, and results are closely linked. Expression is a type of movement-object: a general study of physiognomy in movement is impossible without trying to construct expression. The image, as fixed expression, is both the effect and the instrument of this elaboration. In no case does it refer back to the model, who is like the *provoked* pretext—that is to say, knowingly manipulated for the analysis of expression. Photography is not the image of reality, but one of the means that allow it to be described by rendering it visible.

Significations

In 1864, E. R. A. Serres, member of the Volta Prize Commission, gave his report on *Mécanisme de la physionomie humaine*:

The author opens his book with that sublime phrase from Buffon, "When the soul is agitated, the human face becomes a living tableau where the passions are rendered with as much delicacy as energy. . . ." It is this problem of transcendent physiology, these pathetic signs and these images of our secret agitations which [Duchenne] proposes to produce experimentally and at will by electro-physiological analysis, at the same time as he seeks to fix their expression by photography. The enterprise is lofty and bold; to some extent it is the consequence of modern studies of the nervous system and cerebral functions, whose experimental procedure it perfects, but it does not attain the principle of action of this organic system, which itself is only an instrument. In effect, if a man is an intelligence served by organs, we may conceive how they artificially produce their actions, but left outside the experiment is the principle that makes them act naturally. This principle is inaccessible to the compass of science. It is essential to recognize this from the start, along with all eminent physiologists, when the functions of the cerebral-spinal axis of the nervous system are tackled experimentally. . . . But apart from this general observation—independent of the experimenter since the soul that is their seat and their motive lies outside his reach—apart from this observation, one can only be forcefully struck by the scholarly analysis of this anatomic physiology of the facial muscles and of their movements, so varied and flexible in the expression of the sentiments that may agitate the human soul. These movements are a language, and this universal language is common to all human races in all their varieties. In order to be universal, this language must always be composed of the same signs, and for this to be obtained, this unity of signs must be dependent upon muscular contractions that are always identical. In effect this is what M. Duchesne [*sic*] is demonstrating. . . . Thus as Haller says after Plato, one may admire here the geometry of God, which leaves the geometry and mechanics of man so far behind. One may also admire the thousand nuances of man's physiognomy that the soul puts into effect in its agitations, which are outside the scope of experimentation. . . . It is not, we will admit, without a certain apprehension that we have seen M. Duchesne [*sic*] venture onto such shifting and tormented terrain as psychology in our day. . . . That by localized electrization the reophore has succeeded in assigning with precision one or several muscles to the expression of passions is already a great deal. But what about the expression and explanation of intellectual faculties? And the expression and explanation of human sentiments? Do they not entirely exceed the boundaries of experimental physiology? I would say more: it is to falsify its principle and method to make it enter into the field of metaphysics.[1]

With a mechanistic explanation of intellectual faculties and sentiments, physiology would absorb psychology. With the subordination of mind to matter, physiology would spill over "into the field of metaphysics." By the artificial production of the signs of passions, Duchenne would imply that the mechanism of physiognomy is devoid of spontaneity. Physiognomic movements would be assumed to be the consequence of a simple physiological process. And so Serres was hardening the opposition between natural expressions and deliberate mimicry of them. He was thus stressing the intimate connection that unites natural expression with the data that it translates. If the authentic expression is without equivalent, it is because it is the exteriorization of an emotion: "One may also admire the thousand nuances of man's physiognomy that the soul puts into effect in its agitations, which are outside the scope of experimentation." How movements are commanded by the soul is a problem beyond the reach of electro-physiology: "The enterprise is lofty and bold . . . but it does not attain the principle of action of this organic system, which itself is only an instrument." Serres could not help but reject the experimental part of *Mécanisme de la physionomie humaine.* On the other hand, he retained the part that allowed inverting the relation of subordination by placing the body in a state of dependence on the mind. Only the author of *Anatomie comparée transcendante* [Comparative Transcendent Anatomy] (1859) could identify "this problem of transcendent physiology." Just as language is the instrument by which God reveals eternal ideas and reveals himself to the human spirit, so too the natural language of the passions is the *expression* of the spirit. Nothing prevents the study of the language of the passions, to the extent that it is the analysis of the *signs of a divine language*: "In effect, if a man is an intelligence served by organs, we may conceive how they artificially produce their actions." Serres was reestablishing the transcendence of the intelligence and the interpenetration (*intussusception*) of idea and language. The biological signification of the analysis of passions was turning to the benefit of another metaphysics. God created nature through the Word: the transcendence of ideas requires an intermediary that expresses them to the spirit. By this stroke, mimicked expressions were becoming transcendental forms. With the artificial production of the signs of passion—irreducible to the conventional signs of language—Duchenne had supposedly offered proof that emotional language derives from God and from him alone takes its power of expression. "To be universal, this language must always be composed of the same signs and, for this to be obtained, this unity of signs must be dependent upon muscular contractions that are always identical. In effect this is what M. Duchesne [*sic*] is demonstrating."

Things are not so divinely simple. It will have to be shown how Duchenne wove a series of links between observation and experience. Very schematically, the imbrications of three lines of research define this first psychology. Duchenne constituted a semiology by relying on a symptomatology of the passions, from which he derived his theory of expression. But it was through different modes of the insertion of the subject into his experimental plan that he could envisage a theory of the passions. Electrophysiological analysis, because it was linked to the philosophy of the natural sign, embraced a theory of the natural language of the passions. Starting there, we shall see why the attacks of Duchenne's contemporaries against him were unjustified. The critiques of A. Dechambre, Amédée Latour, and Louis Pierre Gratiolet are of the same nature as those of Serres. For them, an insurmountable barrier separates the domain of physiology from the domain of psychology. Physiology therefore could not give a foundation to a psychology whose supporting walls would consist of a theory of expression and a theory of the passions: a theory of expression was necessary because physiology could not offer more than was given by anatomical deduction (movements and not meanings); a theory of passions was necessary to the extent that physiology could attain the principle of the actions of the emotions. To these criticisms should be added Darwin's: Duchenne was "generally in the right," but that does not mean he is always correct. Someone might be situated in the field of a new discipline and yet formulate false propositions. Darwin thought that Duchenne had committed the same error as most of his contemporaries. In rejecting the principle of evolution, he was regarding each species as the product of a distinct creation. As Darwin quotes Duchenne, one might say that "Duchenne's error" is his fatal reference to the wisdom of the Creator as author of the natural language of the passions. It is not surprising that Darwin contests this metaphysical plane. But was this not for Serres the only good thing, since it came from on high? These criticisms are revealing of the inevitable resistances that any new kind of thinking encounters. Creating a biology of the passions, Duchenne ruled out a series of themes that had arisen from a metaphysics of the senses. Until then, the language of facial expressions depended on Will, Nature, or History. In his effort to establish expressive function as a language of action through the analysis of its syntactical rules and its morphological figures, Duchenne was the first to attempt to understand its mechanism and its significations.

The Language of Passions

Simple or combined muscular actions designate a play of physiognomies, each of which has its principle of composition and rule of appearance. The face is composed of an ensemble of muscles that are each both signifier and signified. Different degrees of expressivity of muscular contractions must no doubt be distinguished: some are harmonious, others imperfect, and the latter assume expressive value when they enter into combination with other contractions. "The experimental study of isolated contractions of the muscles of the face shows that they are either *completely expressive, incompletely expressive, expressive in a complementary way,* or *inexpressive*."[2] All muscular actions are manifestations of a passionate nature and are situated in a field of expressivity. They are inscribed in a scheme that signifies the passions in their expressive actuality. The muscles that manifest emotions, sentiments, and acts of understanding are organs of the expressive function. Something in the immediacy of muscular actions signifies the passion, by which it is opposed to phenomena relating to the inexpressive. In effect, a completely expressive muscle ceases to be so the moment its contraction exceeds a threshold of expressivity. Thus muscular movement tips into forms of the pathological. There is a degree of contraction of the superciliary beyond which this muscle loses its expressive property: "We frequently see this spasm of the superciliary in people who are disturbed by a very bright light. . . . He was contracting his superciliary in a strong and involuntary way."[3] When they are moderate, the simultaneous electrical contractions of both the zygomatic majors and the superciliaries produce an expression where joy and pain are mingled. But when they are energetic, one observes a grimace similar to a convulsive spasm, which doctors call "an indolent facial tic." Finally, there are muscles whose contractions present no expression whatsoever. These muscles, although inexpressive, become expressive when they are associated with the contraction of other muscles. For example, the muscle of the neck lowers the integuments of the lower part of the face and swells the anterior portion of the neck without leaving the least expression. But when its action is associated with the contractions of the frontals, an expression of fright appears. Through this opposition, sometimes in forms of pathology, sometimes in the grimace, and sometimes in the inexpressive, the expressive movement abandons its status as a natural phenomenon and becomes representative of emotion—that is to say, of itself taken as a whole.

Hence the fundamental ambiguity of muscular actions. On the one hand, they refer back to an expressive function and to the link with an ensemble of natural phenomena that together define the language of passions. The laws governing human facial expression may be sought through study of muscular movements. On the other hand, in its expressive function, muscular movement refers back to the difference that separates the expressive from the inexpressive, the discordant, and the pathological. Muscular action signifies the totality of what it is, that is to say, a natural expressive movement and, through its emergence, the exclusion of what it is not: a muscular spasm, a grimace, an insignificant movement. Thus when a muscular action is significant in relation to itself, it is doubly signified: by itself and by the emotion that, in characterizing it, opposes it to inexpressive phenomena. But taken as signified, either in itself (natural language) or through the emotion (in its difference from the inexpressive), it can receive its meaning only from an act that has in advance transformed it into a sign. Psychological thought merely transposes, into the vocabulary of electro-physiological experimentation, a conceptual configuration whose discursive form Étienne Bonnot de Condillac had already laid down. In psychological thinking, muscular movement plays the role of the language of action. It is similarly taken in the general movement of Nature. Its force of manifestation is as primitive as the instinct that carries this initial form of the natural language. In assimilating the superciliary to the "muscle of pain," Duchenne inscribed in its contraction the act of a moved subject enunciating his emotion. But it is through *physiognomic movement* that it provokes what the muscular action simultaneously exteriorizes. There is thus a reduplication of a signification that was *already* accessible to analysis. In perceiving the images of passions that appear on the face, Duchenne hears a language. The folds, wrinkles, and lines of the face are signs of the physiognomy in movement. A second requisite of psychological thought was that in order to insert a structure of expressivity into muscular actions, those actions have to communicate, in their own right, with the signs of the moving physiognomy. At first sight, this is nothing that was not already known. Moreau had identified the signs of passions with muscular actions and physiognomic movements. But he remained at the level of abstract generalities. Expansion, oppression, and convulsion are signs that indicate different classes of passions as well as the appearance of the corresponding synergetic movements. By breaking things down into new objects in experiments, it becomes possible to constitute a semiology of passions, which is founded on the elucidation of correlations between muscular actions and the play of facial expressions.

Bringing expressive muscles to light was inseparably linked to the analysis of the signs of the face in movement. But, inversely, that analysis is inseparable from electro-physiological examination. The study of signs of the face in movement defines what Duchenne calls "the symptomatology of passions." Signs and symptoms are (and say) the same thing, with two small differences: not only does the sign say the same thing as the symptom but, in its organic reality, the sign is *totally* identified with the symptom and with the muscular contraction that provokes it. The symptom, and the muscular action, constitute the indispensable supports of the sign. Consequently there are no signs of the face in movement without symptoms and without muscular actions. But what makes the sign a sign belongs neither to the symptom nor to the muscular movement. It cannot acquire a signification except by the oblique path of the psychologist who first detects its *indices*.

Hence the detachment of the sign from its morphological and physiological supports. Duchenne was here confronted with the problem of the transformation of the symptom (and the muscular action on which it depended) into signifying elements—that is to say, into elements that signify the emotion as the immediate truth of the symptom and of the muscular contraction. In order to resolve this problem, Duchenne made transparent, through methodic experimentation, the whole field of perception. The operation consisted of comparing artificial modeling, which results from the excitation of the motor muscles of the superciliary, with the natural modeling provoked by the sentiment of pain. The superciliary becomes an expressive muscle because it imprints a frown on the eyebrows: the habitual symptom or indication of pain. Moreover, the operation dissipates the appearance of a modification to the whole face by showing that the change is due to the contraction of a single muscle. For expressions requiring the combined action of several muscles, the operation consists of recording the frequency of the simultaneity. What is the relation between the contraction of the zygomatic major and that of the inferior palpebral orbicular? "Experimentation, coupled with observation of naturally expressed movements, has shown me that this particular modeling of the lower lid develops from emotions that agreeably affect the soul and that it completes the expression of smiling and of laughing." Excitation of the zygomatic major, when it is not accompanied by that of the inferior palpebral orbicular, gives a false laugh: "Compare figures 30 and 31, in which the laughter is false and deceptive, with figure 32, which shows this man when I had made him spontaneously laugh; you sense that in the latter his laugh is honest and open."[4]

Duchenne establishes the very first link between the physiology of movements and psychology—or to be precise, a psychology without states of the soul. Semiology is a form of reading verified by the technique of localized electrization. The *anatomy of passions* designates a theory of passions. By the intervention of the electrical analysis in physiognomic perception, Duchenne was distinguishing between artificial expressions and natural expressions. The fabrication of the former illuminates the mechanism of the latter. If quite *naturally* the face in movement speaks the language of passions, it is because the signs of this natural language could also take shape and value within the interrogations posed by electrophysiological investigation. Nothing prevents signs from being solicited or fabricated by this investigation. The sign is no longer just that which, in the felt emotion, is spontaneously enunciated. It also belongs to the point of provoked encounter between an experimental gesture and the experimental subject. To make a sign artificially appear where there is no affect is to arouse a response when the soul is indifferent. The epistemological and perceptual structure that governs electrization of facial muscles is that of the invisible made visible. The face of a person is a surface upon which are registered the expressive lines of the passions, as well as a transparent veil. From the moment when the method prescribed for psychology to interrogate the face at the level of muscles and to make apparent on the surface what is given only underneath, then the technical artifice capable of unleashing a contraction becomes a fertile idea. To establish the signs of passions is to throw across the face a whole network for locating the features founded on muscular movements. Moreau, Sarlandière, and Cruveilhier knew well that all expressions depend on muscular actions. But they stumbled over the physiological analysis, which was now opening the path from muscular actions to expressions and vice versa. Physiology and semiology rely upon each other. Every facial expression envelops a physiological mechanics, a structure of fundamental expressivity.

The whole value of the anatomical study of passions resides in the production of signs that are at once artificial and natural: artificial, because the cause that might produce them is electricity or volition, and natural, because in the eyes of the observer nothing distinguishes them from the spontaneous signs of passions. Distinct in their sources, the two kinds of expression are identical in their visible manifestation: *incarnated* passions. Hence the comparative study of natural, voluntary, and artificial expressions, but also the analysis of their combination. Thanks to deliberate comparisons and associations, expressions end up pronouncing their truth. The comparison might be between natural and artificial expressions. Duchenne attentively followed the gradation of crying expressions in children. He saw tears always foreshadowed by the partial action of the zygomatic minor and full weeping shown by the action of this muscle and the sphincter of the eyelids. It sufficed to electrify these muscles to obtain the same nuances of the expression of crying. Or the comparison might be between voluntary and artificial expressions. A young man called Jules Talrich could make the two superciliaries move together and thereby mime deep pain. Duchenne artificially put these same muscles in action and saw the movements, lines, and contours that he had observed in the mimed expression. Combinations consisted of the association between artificial and voluntary signs. The contraction of the zygomatic minor (artificial) and the eyelid sphincter (voluntary) gives one mode of crying: full weeping. By the contraction of the elevator specific to the upper lip (artificial) and the eyelids (voluntary), one obtains a nuance of the same crying: the subject seems to abandon himself completely to his emotion. But associations may also mingle natural, artificial, and voluntary signs. The contraction of the triangular of the lips (natural), the transversal of the nose (artificial) and the palpebrals (voluntary) gives the sought-after expression: "I profited . . . from the lowering of his labial corners to study the expressive effect of the combination of the nose transversal with the lip triangular. The signs of discontent did appear . . . but this expression was not yet the exact imitation of nature. I then had him close his eyelids, as if he were bothered by the light . . . and I saw appear on his face, with a perfect true, a mixture of discontent and contempt."

Whatever the mode of producing expressions, it is enough for the same muscles to enter into play, to see that the signs produced are the same as well. The experimental production of signs of passions coincides with the symptomatology of passions. What is seen after the electrization of a muscle is identical with what is perceived under normal conditions in real life. It matters little that the experimental subject feels no emotion. It is enough that the passions are rendered in conformity with the laws of muscular physiology. Duchenne's goal was to make each expression an organic event. The space of his analysis is that of absolute visibility. Through relating a perceptive sector to a semantic element, the visible becomes readable. Of course, the expression of a passion without a mood is not an authentic element of passion: the discourse of the surface is opposed to the emptiness of mood. But this discourse coincides with the natural language of passions. And the artificial passion seized by the lens is put into relation with the experimental subject's frame of mind. Thus Duchenne traces a line of division between the expressive surface and the empty interior. The subject's state of consciousness is a profound apathy. During the experiments, the mind of the old man is alien from what is read on his face. The provoked action of his zygomatic muscle makes him break out in an honest laugh; that of his superciliary contracts it in intense pain. Yet inside him lies neither any gaiety nor any painful preoccupation. Not only does he feel nothing, but he would even be incapable of feeling the passion being read on his face: "The old man shown in figure 39 is far from lascivious; he is on the contrary of such a cold temperament he vows that women have never inspired him in the least. He is even proud to have conserved his innocence. Moreover, we notice in his photographic portrait, from the shape and flattening of his nostrils . . . , it is clear that the expressive muscle of lascivious pleasure is in his case poorly developed."[5]

Muscular actions designate a physiological determinism that accounts for expressive modeling. Under the skin lies a *plastic* mold, or to be more precise, a matrix where there is crudely struck the imprint of *character*: the signs of passions. From the equivalence between the artificial and the natural results the elucidation of the functions of mood. Source of expressions, this is what puts muscles into play and makes them depict on the face, in characteristic features, the image of our passions. Each facial (and dependent muscle) movement is surcharged by the transparency of a passion with the syntax of a descriptive language. Hence the fundamental isomorphism of the signifying structure of emotion, and the verbal form that surrounds it. The descriptive act is, in its own right, a description of an essence. But inversely the latter is not given, in a double physiological and physiognomic movement, without offering itself to the mastery of a language that is the very speech of passion. The psychologist's gaze is able to understand a language at the moment when he perceives a spectacle. It could not be otherwise, given the essential relation between the perceptive act and the element of language. Duchenne could thus make physiognomy speak. The gap between the experimentation that activates the muscles and the expressions that come onto the face corresponds to the difference between the question that is posed and the language that is heard: "It is a problem for which I sought a solution for many years, provoking with the aid of electrical currents the contraction of facial muscles, to make them *speak* the language of passions and sentiments." It is not surprising that Duchenne gives electro-physiological exploration the authority of Francis Bacon: "the experiment is a sort of question applied to nature to make it speak."[6]

Experimentation interrogates in the vocabulary of nature, inside the language that has been proposed to it by things observed. But its questions can only have a foundation upon condition of soliciting responses that concur with the language that is the principle and law of the composition of expressive forms. Electro-physiological analysis and psychological perception deconstruct in order to bring to light a prescription that is the natural order of signifying muscular actions. Moving from the simple to the composite and acting alternatively on one side or on both sides, Duchenne made muscles contract two by two, then three by three, varying the muscular combinations. Through this analytical gaze, the perceived and perceptible could be reconstructed in a language whose rigorous form enunciates its origin: "Some of these original expressions, as we have seen, are perfectly drawn by partial contractions of certain muscles, whereas other original expressions that individually are also represented by a special muscle have need, in order to be complete, of the cooperation of one or more other muscles."[7] On the one hand, the psychologist's gaze is defined as a perceptual act underpinned by a logic of operations. It is an analytical gaze that reconstitutes the genesis of composition. On the other hand, observation is confronted with the linguistic structure of expressive forms: the invention of biological expressivity amounts to investing the movements of the face with the forms of language. But by the analysis, which is the act of reconstructing the original expressive forms, Duchenne was touching an insurmountable barrier, a threshold he could translate only in terms of predetermination. The language of passions arose from the play of muscular actions instituted by an ingenious Nature: "These experiments, by making us penetrate the mysteries of physiognomy, prove (if that still needed to be demonstrated) that it is not the product of chance; they prove . . . that these composed contractions, so knowingly calculated (so to speak), give a material form to our most secret emotions, that these muscular contractions must be the work of a divine intelligence."[8]

Each passion had its own muscle, its muscle of expression, and so a preliminary table could be compiled of binary correspondences between muscles and their expressive signification. Each muscle could thus receive, alongside its anatomical name, a synonymous denomination indicating its expressive function. This representation is not revealed by the table itself, but had preceded it, since the correlation between each muscle and its expressive value appeared at the point of intersection of the electrophysiological and symptomatological analyses. Inversely, in a second table, the different expressions are attached to the muscles that produce them. This is why Duchenne contrasts his classification with the arbitrary listings of passions found in the writings of moralists and philosophers. Thus he has devised a classification system that approaches the natural order, in which he proceeds to an inventory that is well founded: "I will not be accused of having numbered them arbitrarily, for they are the reproduction of the one the soul itself depicts on the human face."[9]

In the old lineage that still relates the grammar of passion to tradition, Duchenne's classification reproduces the natural hierarchy of passions. An order of decreasing dignity intersects with an order of increasing complexity. The new classification of passions, undoubtedly psychological, fits within Plato's metaphysics. The noblest expressions occupy the upper part of the face. First are the frontal, the muscle of attention, and especially the superior palpebral orbicular, the muscle of reflection: "*Intellectual reflection*, the most important, most noble, and most abstract state of the spirit, and *meditation*, which is the mother of great ideas, are the dominant passions in certain men."[10] Then come muscles whose actions express feelings like pain, aggression, anger, joy, laughter, and weeping (superciliary, pyramidal, major and minor zygomatics, the elevator specific to the upper lip and the elevator common to the nostril and upper lip). Finally, there are muscles that are most often associated with the most bestial expressions (triangular of the lips, square of the chin, the muscle of the neck).

Claude Bernard, who wanted man to become "a veritable foreman of creation," was not wrong about this. *Mécanisme de la physionomie humaine* shows that facial muscles and nerves are mechanical apparatuses created by the organism and governed by laws. But he added a consideration that seemed to offer the highest interest for general physiology and psychology: "M. Duchenne was able, in acting on various face muscles whose action he had determined, to obtain successively on one and the same face the manifestation of various and opposite passions that the model was not feeling at all. Even on a recently dead cadaver, he was able to produce the appearance of similar expressions by electralizing the same muscles."[11] At first sight, there is something paradoxical in experimental demonstrations that underlie the relation between physiology and psychology, in particular those that caught Bernard's attention: the mortuary mask awakens, features are animated; an expressive smile, a poignant pain, an intense scorn, an admiring attention, an aggressive threat—all appear at the will of the experimenter. We know that Duchenne had given up photographing his experiments on cadavers, but it suffices for him to have done so in order to grasp the purified form of his biological philosophy. He showed that muscular actions are the elements of expression and that Nature works with very simple means. It is enough to mention these morbid experiments to understand, if not justify, the worry of his contemporaries. How could you apprehend life within death? How could you grasp the trembling of passions in an experimental subject who felt nothing? Gratiolet reproached Duchenne for missing the specificity of the emotional register, which was describing an order of phenomena that could not pretend to the dignity of a biological function. Affectivity is quite mechanical and artificial; the subject remains alien to his expression and the masks are empty of any emotive content. "A doctor who is justly famous recently believed he had solved the mystery of the physiognomic language by artificially producing movements with the aid of certain electrical currents very expertly directed. These movements may, in truth, simulate expressions; but are they veritable expressions? . . . To produce an expression, to determine with more precision the muscles whose contraction modifies the form of the face—does this mean knowing the true principle behind, and the primary reason for, these movements?"[12] It was easier for critics to denounce the manufacture of emotions than to understand the rules for the fabrication of expressive mechanisms.

Psychology and Physiology

Most of Duchenne's contemporaries did not see that he was opening up a new realm: the functional study of behavior. His subject was not limited to the mechanism of muscular movements; it extended to the face as a field of expression. In 1899, Édouard Brissaud reactivated all the criticisms formulated forty years earlier by Serres, Gratiolet, Albert Lemoine, Dechambre, and Latour: "Duchenne tried electro-physiological analysis of the expression of passions, and he thought he was seeing physiognomies where there was merely an *artificial spasm* of the face. . . . In spite of this reservation, there is a lot to be admired. Electro-physiological analysis, not of the expression of passions but of the mechanism of facial movements, is a pure experimental masterpiece."[13] This was not innocent praise. From the physiological standpoint, nobody was contesting that the experiments fully attained their objective. On the other hand, with respect to psychology, Duchenne was said to have totally failed. In recording surface effects, he had missed deeper significations. It is useful to go back over the methodological presuppositions on which all this criticism is based. Psychology is a science distinct from physiology: "It is impossible for these two sciences to be confused because in the alliance of the soul and the body each conserves its nature and its destiny." But psychology should meet up with physiology, "nor can they forever march separately and independently of each other, because despite the diversity in their essence, the body and the soul are closely united and interdependent."[14] This relation between soul and body might be apprehended in two different ways: either in terms of causality, where the soul is the source of physiognomic movements, or in the mode of a correlation, where facial expressions are the translation of the soul's emotions. In sum, there is no theory of expression without a motive and no theory of passions without affects. But the motive for passions, and passion as a state of the soul, did not seem within the scope of localized electrization.

Analysis of physiognomic mechanisms gave grounds for an initial criticism. Study of muscular actions differs from that of expression as physiology differs from psychology. The action of the electrode cannot give to the face any more than it gives to the shoulder: a change of form and a movement in relation to the muscle put into contraction. Just as Duchenne had determined the role of the deltoid in the action of lifting the arm, so he had determined the role of the triangular of the lips in the movement of the mouth. "But the contraction of this muscle, says M. Duchenne, is the sign of sadness! Who can be the judge? Merely taste, and no longer physiology." It is legitimate to record a contraction and the movement that results, but the physiologist's knowledge stops there. Physiology examines displacements for which one cannot deduce the slightest signification. The study of signs of passions does not fall within its domain: "The truth of a passionate expression, although expressing itself by anatomical means, does not derive from anatomy, but from empirical observation and from art." On this point, each person is an expert. Dechambre gives an example: "In granting that each of the passions mentioned by M. Duchenne has its principal sign in the contraction of one or another muscle, I still cannot accept that the isolated contraction of this muscle—to no matter what degree—necessarily expresses a determined passion, and that one only." The elevation and bringing together of the heads of the eyebrows renders pain. But these movements might express surprise, or mockery, by their alliance with other expressive signs of the face. "Each person can make an experiment in front of a mirror. This is not a painful expression that is tempered or modified by another expression:

the former is absolutely lacking in all the degrees of the contraction of the superciliaries." Dechambre also revives a theme prominent in the *Encyclopédie*: questions of expression pertain to the competence of painters and sculptors, and so it is up to artists to inquire into the theory of expression. They may treat this subject because they are more aware than anyone of "the ordinary complexity of expressions that artistic fictions (richer than electro-physiological experiments) can impose on a single face, and even (not leaving physiology behind) the diverse signification that the contraction of a single muscle may assume, according to the force of this contraction, or at the behest of synergetic actions."

Curiously, though, Dechambre was not challenging the role of anatomical physiology. To represent expressions well, it was indispensable to understand their physiological mechanism: "Whether passions are simple or composite; whether they are calm or violent, pleasant or sad, what matters to us is that they are rendered naturally, that is to say, in conformity with anatomical facts and the muscles' normal play, which are the mechanical means of expression."[15] Thus there is ambivalence, even inconsequence, to his criticism. Dechambre invokes empirical observation and taste as the sole criteria, but he insists on the necessity of grounding the rendering of emotions in muscular physiology. He does not see that the natural is the endpoint of a purpose and that the muscles' normal play designates a fact just as much as a value. Dechambre could not understand how Duchenne had been able to slip his physiology into the heart of the analysis of expressions, nor how he coordinated physiological experimentation with observation of passions. Duchenne's physiology went much farther than simple anatomical deduction. It was a physiology of expressive muscles because it relied on knowledge of the normal functioning of physiological mechanisms. Duchenne established a theory of expression on the model of medical semiologies, but its structure was symmetrical and converse to them. In the clinic, the symptom is a natural phenomenon that also signifies the pathological, in contrast to a natural phenomenon arising from organic life in a normal state. By this simple opposition to healthy forms, the symptom becomes signifying of illness and is associated with an underlying lesion. Conversely, in the symptomatology of the passions, the symptom is a natural phenomenon, since its emergence is linked to its law of appearance. But what in the immediacy of the symptom signifies the expressive is the opposite of the pathological. By this simple contrast with the forms of the pathological, the symptom and its muscular double become signifying of the emotion. The psychology of passions operates in a field structured by the opposition between the normal and the pathological. Symptomatology, which allows indices to be identified, is intimately linked to physiology. To challenge this new theory of expression, it was necessary to occult the central problem of transforming physiognomic movements and muscular actions into signs of passions.

Moreover, Duchenne had already dismissed the criteria invoked by Dechambre: the spontaneous observation of passions and the judgment of taste. The action of orofacial muscles is scarcely different from any other muscular movement. The real mistake to be avoided is not aligning the electrization of facial muscles with that of the shoulder muscles, but rather not seeing that it gives the same thing in both cases. The action of the reophore on the deltoid produces a deformation, an unsightly movement identical to the deformity that is observed in individuals suffering from atrophy of the major serrate. By contrast, the double electrization of the deltoid and the major serrate in a healthy subject gives a natural movement of elevation of the arm conforming to laws of synergies—"natural" here meaning a harmonious movement that even the most gifted artist does not always perceive—perhaps an allusion to El Greco's painting: "If I were dealing with an artistic study of certain canvases that in other respects are masterpieces, I would show shoulder blades that move away from the thorax like a wing during the raising of the arm, a deformity that is the sign of the paralysis of the major serrate!"[16] With the example of disdain, the distance that separates scientific knowledge from common awareness can be readily seen. Anyone can say that lifting the shoulders expresses disdain, but only Duchenne could specify that it is the median portion of the trapezoid, by directly raising the shoulder, that expresses the disdainful gesture. This precision presupposes discovery of the use of the median portion of the trapezoid to maintain the shoulder stump at a normal height by its tonic contraction and to contribute to the vertical elevation of the arm. It also presupposes the elucidation, by clinical study of the atrophy of the major serrate, of the action of the middle third of the trapezoid in the movement of vertically raising the arm. In this movement, the action of this part cannot supplement that of the major serrate. But it can produce the direct elevation of the shoulder stump. In pinpointing the latter movement, Duchenne could attribute an expressive signification to it.

The second criticism was aimed at the theory of passions. Electrization of a muscle produces a muscular action that lays out the signs of an emotion. But, asked Gratiolet, "how can movements communicated to my muscles by a foreign will recount my own sentiments and desires? They could only express the experimenter's idea, fashioning me as a sculptor fashions clay."[17] In founding his theory of passions on physiological experimentation, Duchenne seemed to have forgotten that the subject of psychology is indivisible. He was not describing an incorporeal reality and its correlate, natural expression; instead, he was fabricating simulacra. On one hand, the failure of electro-physiological analysis was stressed; it was powerless to render the effects of true emotion. The play of facial features when responding to a mood offers nuances that are infinitely more varied, more delicate, and more expressive than those of artificial expressions. In passing, Dechambre indicated the evident limit of physiological experimentation: it could not reach the eye muscles. "On this living and mobile scene of facial expression, the eye (that escapes the reophore test) spreads the light and shadow that transform the signification of the muscular contractions and make the gaze, as the poet said, the daylight of thought."[18] On the other hand, photographs were disparaged since they magnified these defects—hence the criticism from Brissaud: "How could Duchenne not have remembered that almost all the true physiognomy is in the gaze, and that we say (literally and not metaphorically) 'soft eyes,' 'sad eyes,' 'angry eyes,' 'sympathetic eyes,' 'jealous eyes'? . . . Duchenne's personage, on the contrary, has fixed eyes, invariably fastened on the lens, and in truth these eyes are without expression; they might have no pupils, like the eyes of a blind person whose eyeball is totally white, or like those of ancient busts that seem neither to look at nor see anything."[19] Finally, Serres thought he was delivering a decisive blow by saying that Duchenne's images did not render the truth of natural expressions: "There is still no doubt something stiff and cold in these artificial expressions; the life is absent, which (we have to say) sometimes gives the air of caricature to the human face thus put into action."[20]

Duchenne's adversaries moved without any problem from psychology to aesthetics. They did not perceive the evident difference between presentation of a function of expression and presentation of plastic expression. Their experience was identified with the exercise of sensitivity alone. As offered by the face, truth is sensitive—for example, the power of the gaze in the work of the passions. This kind of sensitive knowledge, which is skill without concept, falls back on aesthetics. The true and natural physiognomy is the *made* physiognomy. The truth of expression resides in measurement: grace defines its authenticity, relative to which the facial expressions presented by Duchenne were perceived as artificial. From the standpoint of aesthetic perception, his images appeared as figures of bad taste. This perception is remarkably coherent. If the presentation of passions arises from a judgment of taste, then Duchenne is understood as an artist and not a physiologist: for Gratiolet, he is a photographer who works like a sculptor shaping clay; for Dechambre and Brissaud, a bad painter, since he transposes into imagery the classical procedures of antique sculptors; for Serres, a caricaturist, which in this case is not praise. Duchenne's detractors did not see that he was aiming at biological function: the normal mechanism of facial expressions. Nor did they understand how he could present his theory of passions in the form not of a series of expressive portraits, but in an album of expressive figures. The principle of intelligibility of images was at the opposite pole from that aesthetic sense that they so naively played off against it. In the plastic arts, expression should be true to life while revealing the beauty of the subject. Rendering expression tried for an effect that was obtained by erasing both the morphological imperfections and the remnants of expressive features. By this double opposition to natural forms, the expression became a signifier of an emotional state. By contrast, in the scientific analysis of passions, signs are presented such as they appear: without dressing up or overshadowing an unattractive physique. That which in a fixed expression signifies passion is opposed to a conventional image. The signs are natural and they reflect what they designate: a function in action that only photography can represent. By this simple opposition to the forms of *presentable* expression, the indices of passions are associated with the normal functioning of the muscular apparatus. Duchenne operates in a field structured by the opposition between the normal and the pathological, but also by the opposition between the natural and the conventional.

Incidentally, the doctor of Boulogne had precluded an improvised phenomenology and the privileges of sensitivity. The observer is always confronted with phenomena whose perception is received as conforming to the observed object. One would be wrong to think that the eye receives, reflects, or spreads an interior light, whether the daylight of thought or the warmth of sentiment. The gaze is the mirror of the soul because the face receives and reflects the light of day and its shadows. Modifications of the gaze depend on the remnants of what surrounds it: the eye takes on the color of passion under the effect of muscular actions. The gaze becomes interrogative with the elevation of the frontals, menacing with the action of the pyramidals, and smiling with the contraction of the zygomatics major and the inferior palpebral orbiculars. Moreover, Duchenne had not omitted to indicate the direction of the eyeball axis in certain passions: for example, an oblique gaze upward and sideways for ecstasy and sensual delirium, a downward gaze for humility and sadness. But the most serious error of his opponents was to not have understood that the photographs are inscribed only within the scientific register. Serres is a man of taste, and he judges Duchenne's images on the basis of the rules of art. In photos that fix the acme of a passion, he was seeing prototypes close to caricature. He did not understand that these images escape the genres of both portrait and caricature. Duchenne did not want to either attenuate or correct facial features, on the one hand, or to deform or emphasize one feature or another. Serres's plastic references led him to contrast the readability without excess of the portrait with the excessive readability of the caricature. He did not understand that Duchenne's aim was not to please an art lover but to instruct his reader. And Duchenne knew how to do so. In any case, it was not he who would have modified, in the expression of sadness, one of its most troubling traits: the contraction of the zygomatic minor, which gives a person a silly and ridiculous air when he cries.

Darwin, finally, did not hesitate to stress the error he thought Duchenne had committed, supported by this quotation: "The Creator has not had to concern himself here with mechanical requirements; he has been able, in his wisdom, or—if you will forgive the expression—by divine whim, to move one muscle or another, one or many muscles at a time, when he has wanted the characteristic evidence of emotions, however fleeting, to be written in passing on the face of man."[21] We understand why Darwin did not want to hear the wisdom of the Creator mentioned; did this not touch on one of the essential points of the theory of evolution? In *The Origin of Species*, Darwin stressed the "unity of type": "What can be more curious than that the hand of a man, formed for grasping, that of a mole for digging, the leg of the horse, the paddle of the porpoise, and the wing of the bat, should all be constructed on the same pattern." There were two irreconcilable explanations. On one side, Darwin's "theory of descent with slow and slight successive modifications," and on the other, the doctrine of ultimate causality, the ordinary view of the independent creation of each species: "that it has pleased the Creator to construct all the animals and plants in each great class on a uniform plan."[22] From the moment Darwin excluded the principle of purposefulness, he no longer saw that the determination of a goal might remain a legitimate requirement of physiology. Galen had spoken of the mechanical system of the human body as a mirror that reflected the skill of the Architect. As soon as one identifies dispositions that assure the production of a result, the idea of an Intelligent Designer becomes apparent. At a time when Charles Daremberg was editing and commenting on the works of the Pergamum doctor, Duchenne could not help being seduced by this idea. Incidentally, no other domain than anatomy was as welcoming to the principle of purposefulness. For Duchenne, if "the Creator was not concerned with mechanical necessity," it was because he was concerned elsewhere: with the synergetic movements governed by the laws of mechanics. In 1853, he wrote: "As Galen, the profound author of *On the Natural Faculties*, would have said, we see how clever and prescient Nature has been in giving the diaphragm the faculty of rendering to the thoracic cavity in breadth more or less what it loses in height."[23] Thus the alternative was the following: either any reference to divine intelligence is the mark of spiritualism and the whole physiology of movements tips over into metaphysics, or else this theme does not exclude a discourse that it would be presumptuous not to recognize as a scientific discourse. Darwin would have done better to take Duchenne at his word: one had to *pardonner sa manière de parler*, "forgive his manner of speaking."

The Creator is the Cartesian equivalent of a fabricating God: the efficient cause of any bodily mechanics. Whether human or *animal*, signs appear at its mechanical boundaries. A tiny event, reported by Darwin precisely, bears direct testimony: "Doctor Duchenne—and I cannot quote a better authority—informs me that he kept a very tame monkey in his house for over a year; and when he gave it during meal-times some choice delicacy, he observed that the corners of its mouth were slightly raised; thus an expression of satisfaction, partaking of the nature of an incipient smile, and resembling that often seen on the face of a man, could be plainly perceived in this animal."[24] If Duchenne wanted to establish a barrier between humans and animals, how would he have spoken of the monkey? Perhaps haughtily, like Gratiolet: "The smile is impossible for this mouth, the lip and chin are mingled in a sort of rounded valve opposite the upper lip; and when the mouth is closed, their borders, intimately adjusted, are straight, flat, and let no part of the mucous bloom; one feels right away that these lips will never speak [lively applause]."[25] Duchenne was so little preoccupied with high metaphysics that he communicated his observation to Darwin, as well as others relating to the elevation and lowering of the eyebrows in the monkey: movements that testify respectively to great attention and to thinking. The work of the Boulogne doctor, far from blocking a history of emotional language, offers some principal conditions of its possibility. It is pointless to insist on Darwin's debt to Duchenne. The function of expression leads to new relations between the different elements that constitute it, particularly correspondence and interdependence. "Correspondence" means that expressions refer back to the action of specific muscles—hence the "interdependence" of the expressive function and the muscular system. From this flows the identity of modes of expression among all human populations. Darwin could open the field of biological anthropology by launching a major inquiry into the expression of emotions. Finally, by identifying muscles over which the will had no control, Duchenne opened the way to the thesis of innate language. Based on reflex movements, Darwin could put into play the heredity of acquired characteristics and tackle the problem of the genesis of expressions. For the first time, history was introduced into study of the emotional language of living beings.

Discussions within the Volta Prize Commission pitted the few partisans of Duchenne against his opponents. Serres, although reserved, said that the study of the mechanism of expressions amounted to a new contribution. Michel-Eugène Chevreul, no doubt struck by the study of optical illusions, saw it is a phenomenon of the same order as what he had demonstrated in his *De la loi du contraste simultané des couleurs* [On the Law of Simultaneous Color Contrast] (1839). But Pierre François-Olive Rayer, who cast doubt on the originality of Duchenne's work, asserted that Lavater had already established the original principle of his research. Antoine César Becquerel, for his part, contested the novelty of Duchenne's work by evoking Giovanni Aldini's experiments on animal heads. But the blindness and conservatism of the Volta Commission members scarcely count for much in this history. In 1851, in the third edition of the *Traité d'anatomie descriptive*,

Cruveilhier was affirming the existence of the fronto-orbital muscle; hence his embarrassment when he tackled the question of expressive significations: "The soul's hesitation between various sentiments is thus betrayed on this small fronto-superciliary region, which observers and painters cannot study too much."[26] But in the fourth edition (1862–1867), following and according to Duchenne, things were clear: "The frontal is the muscle of *attention*. . . . The contraction of the pyramidal *gives hardness* to the look and announces *aggression*, meanness, hatred. . . . This portion of the orbicular is the muscle of *reflection*; when it is violent, it indicates a mediation with effort, a laborious endeavor of thought. . . . The superciliary is the muscle of *pain*, of *suffering*."[27] At the same time, Hubert de Verneuil was reviewing *Mécanisme de la physionomie humaine* favorably. Along with Claude Bernard, he was the one who best understood the significance and bearing of Duchenne's work. Addressed to psychologists, who were lagging behind a revolution, here is an informed doctor's judgment: "Physiology, the noblest and indisputably the most difficult of the sciences of observation, is now undergoing a complete metamorphosis. . . . In this movement, which one could only compare to the revolution in the physical and chemical sciences at the end of the last century, we must especially note the tendency to restore to physiology its full rights and natural frontiers. . . . The magnificent work of M. Duchenne (of Boulogne) belongs precisely to this extreme frontier where physiology fuses with psychology and the plastic arts, which explains why the subject has been little or imperfectly explored until today by pure scholars. This research forms an important paragraph in the grand chapter of the *functions of expression*."[28]

Biology and Anthropology

One has to start with the function of expression and the postulate that confers its whole meaning: the linguistic nature of muscular actions. Duchenne is close to Condillac on this: the language of the passions is a form of the language of action. It is thus impossible for this immediate language to assume meaning for the gaze of another, if a person does not already possess the faculties of signifying and understanding: "To express and to monitor the signs of facial expression seem to me to be inseparable faculties that man must possess at birth."[29] Condillac had seen this initial form of language as only the effects and result of the conformation of our organs. At its beginnings, this language consisted only of grimaces, cries, and contortions. But such actions are not a language, or even a sign. Without observation and analysis of the relations between his own facial movements and his passions, a person would not understand the meaning of movements made in front of him. And so he would not be capable of explicitly making similar ones to make himself understood. It was necessary for the subject to associate the facial expression perceived on the face of another with the same passion that has several times accompanied his own expressions. He could then receive this expression as the mark and substitute for the thought of another person—that is to say, as a sign. The language of action indeed links, by a genesis, language to nature, but in order to detach it and mark its difference from grimaces: "This language is thus not so natural that one does not know it without having learned it."[30] On this point, Duchenne departs from Condillac: the body, from the start, speaks this language of action. This means that a person receives from nature not the wherewithal to make signs but the signs themselves. The language of passions is not learned. A new line of division is thus traced that no longer separates a prelinguistic material given by nature from the signs of representation, but instead separates the expressive and the inexpressive: the signifying movement and the grimace. To the genesis that linked language to nature in order to ground its artifice in nature, Duchenne contrasts a static genesis. The skin and its muscular double are the right side and the flip side where the opposition between sense and nonsense is manifested. Grimaces are not the material given for the elaboration of signs, but the *insignificant* residues. Not totally, however: in medical semiology, grimaces are apprehended as indices of a dysfunction. Forms of the pathological at the limits of the body are at once inexpressive and signs of a functional disorder. The inexpressive is correlative of the *depressed*. Essentially, Duchenne moves by a notch the anchor point for the *expressed*, to simple or combined muscular contractions. He sees the language of passions as a language of action that is the *analogon* of language, a natural language that is embodied in the muscular system.

Moreau had offered a physiological analysis of passions that interwove an iconographic description of passions and a typology of the emotions founded on the pathogenesis of the soul's illnesses. The former located figures of passion in history painting, while the latter registered the general appearance of movements of the muscles and of the face: the signs of passions. The two modes of decoding more or less overlapped, but they were independent of each other. The made, or civilized, physiognomy responded to aesthetic criteria and to social conventions. The signs of passions had a marked relation with all that comes from art, education, and culture. But facial play was also subject to the classification of the face's muscular maladies: the expression of passions was related to pathological expressiveness. The passionate man, like the sick man, was quite whole in his features, so it was necessary to go back to that point of imaginary contact where the primitive rhythm of irritability touches temperament. The exaltation of contractility, because it is linked to a sanguine temperament, expresses cheerful passions. The tightening of muscles, specific to a melancholic temperament, gives sad passions. The over-excitation of contractility, which depends on a bilious temperament, marks the convulsive passions. The adjustment between an independent iconographic description and a semiological reading had defined the framework for the analysis of expression. This arrangement of knowledge was what Duchenne overturned. By being anchored in the organism, the language of passions designates a network of significations that relates to a *reflexology*. In subjecting facial movements to the mechanisms of muscular actions, expression becomes the element of a function. Duchenne had not only given a cartography of affects but had sought among physiological mechanisms the foundation of all physiognomic movements. Physiologists had talked about the sensitive field and the motor field, and now an expressive field was appearing. What makes expression possible and makes it significant is a series of organic movements. Duchenne "saw" the nerve filaments contracting the muscles and the characters of passions being drawn on the skin. By replacing the passions on the distribution matrix of the roots of peripheral nerves immersed in the muscular bundles, he apprehended the radical signs of expression. But the play of simple or combined muscular contractions is automatic and requires no kind of learning or skill.

The consequences that result from his scheme allow us to measure the distance that separates Duchenne from his predecessors. As long as passions were being analyzed by their figuration, the language of physiognomy crept into the cultural forms of expression. As long as the signs of passion were founded on an analogy with the muscular complaints, the state of contractility was allied with the temperament of the subject. This ensemble of determinations, innate and acquired, encouraged apprehending the diversity of complexions and physiognomic movements according to race, geographical location, and state of civilization. It legitimated the perception of an irreducible variety of facial expressions in order to express the same passion. In 1820, Julien-Joseph Virey could write: "The ferocious anger of a Tartar or an Iroquois cannibal is portrayed by more hideous and frightening traits than the proud and noble wrath of a civilized European who is outraged."[31] From the moment when the language of passions leaves biology behind, then these themes of an emerging anthropology tip into fallacies. To see natural differences in modes of expression arises from fantasy, as Duchenne realized: "If it had been otherwise, the language of facial expression would have had the same fate as that of spoken languages created by man: each region, each province would have its way of portraying emotions on the face; perhaps this caprice would have gone so far as to vary infinitely facial expression in each town and each individual." And to see natural differences in the expressive mobility of the face equally spills into phantasms, whereas actually "the patterns of expression of the human face . . . are the same in all people, in savages and civilized nations." Duchenne identifies the function of expression with a natural language that all humans speak and understand in the same way. This identification is the result of a double yet contradictory operation: the transformation of indices into signs and the naturalization of signs. The transformation of indices into signs because an index is only a spontaneous manifestation of a disposition or of an affection whose actualization is inseparable from a lived situation. In order for the play of indices to end up defining a language, they had to become signs. The naturalization of signs, insofar as they are meant to escape the linguistic structure arising from a voluntary act, is attached to an intention to communicate. From this derives the idea of an expressive language, immediate and involuntary, that no arbitrary power can obliterate. Then, and only then, to express is, in Valéry's words, "to touch the essence of ordinary language." Derived directly from the organization of the body, in close relation with the disposition and play of the muscular system, the characters of expression form an immutable and universal language.

The second postulate complements the preceding one and assures the biological basis of the expressive function. The language of passions derives from instinct: "The language of facial expression had to be immutable, a condition without which it could not be universal. Because of this the Creator placed facial expression under the control of instinctive muscular contractions."[32] Natural language is closely linked to man's destiny. This is why Duchenne places it among the original aptitudes that define human nature. In walking, the child executes instinctively movements of great complexity without the participation of his will; similarly, all the movements of the face are executed with remarkable consistency. It is always the same muscular contractions that naturally portray each passion on the human face. The language of affects is opposed to spoken language, as natural language is to conventional language. The language of passion defines certain actions: it expresses what one feels, what one fears and what one is undergoing. It is close to the subject in his living activity: the passion of the soul serves the individual and contributes to his conservation. But the expressive function also manifests itself when words are lacking and when the body takes the initiative. The language of emotions thus acquires its own expressive value: it manifests the fundamental impulses of those who speak it. The term "commotion," referring both to a physical jolt and an agitated mood, justly highlights the motor element included in the emotion. There is an opposition between the natural language of the passions and the mimicking language founded on the concerted usage of signs. Duchenne sees this usage as merely a body language of significant facial expressions. The situational context enables the meaning to be immediately grasped: expressions replace words and their meaning is implicit. Facial expressions, just like nonverbal language, lean toward the side of spoken language. Duchenne perceives here a secondary, derivative use, which should not supplant the original language of passions nor mask the specificity of the field of expression. In the general economy of his analysis, the exclusion of mimicry is only the negative counterpart of a positive and subtler element: the idea of a field of expressivity that is not used by the subject to say something but that has from the start a vital signification. Thereby expression can designate only an involuntary and immediate action that lies beyond the subject's control.

For almost two centuries, Descartes had fixed the way to see facial expressions. In the facial movements that accompany passions, he distinguished natural movements from intentional movements. Moreau had reactivated this distinction but based it on a division between organic life and relational life. He thereby distanced himself from Bichat, who attributed passions to the organic life alone, whereas Moreau thought sensitivity, voluntary motor skills, and intellectual operations arose from the relational life. He proceeded therefore to a new division of the orofacial muscles: the muscles of the "*face*" belong to organic life and the muscles of the "*visage*" to the relational life. The movements of the face are involuntary. They are linked to instinct and designate the cruel passions that put into play the uncontrolled contractions of the masseters, the buccinators, and the temporals. On the other hand, the movements of the visage come under the relational life. They present the signs of passions linked to sentiment and to intelligence, and respond to the deliberate determinations of the will. Moreau thus rediscovered the Cartesian theme of an expressive language that serves to communicate. Bell's work marked a turning point, since for the first time all physiognomic movements become "symptoms." They arise spontaneously and are carried out without cooperation from the will. The very complex synergies of the respiratory function are assured by the face's motor nerve, which becomes the respiratory nerve of the face: the agent of the movements of facial expression. Although the heart is not the seat of passions, it is under the influence of moods. From cardiac disturbances result modifications of the respiratory function that make their effects felt on the throat, lips, and nose. But Bell's demonstration rests on study of the relations between movements of facial expression and of respiration in extreme situations: during performances, agony, and pathological states that occur at the height of respiratory difficulty. Duchenne suppresses all these mediations by making the involuntary action of facial muscles depend solely on the seventh pair. From there, a symptomatology of passions with regard to a person agitated by emotions under normal living conditions was possible, which did not preclude, through self-imitation, some of these movements being assumed on a voluntary level.

We again see the immense distance that separates Duchenne from his predecessors. With the assimilation of facial movements to reflex actions, he overturns one of the principal obstacles that had inhibited the physiology of passions since Descartes. As long as it was thought that facial expressions were commanded by the will, something like an expressive function was unthinkable. Duchenne did not doubt for an instant that man could simulate passions. He knew that the face could be the instrument of a speechless language, founded on the will and the utilization of conventional signs. What he contested, though, was that the soul uses facial actions solely in order to declare (or dissimulate) its passions. In ruining pathognomy, Descartes had gone too far, since he has also precluded a theory of the language of affects. For him, laughter would never become the sign of joy; similarly, no facial expression would indicate anger for certain. We have to return briefly to *Traité des passions de l'âme*—more precisely, to operations of the will that sometimes use natural signs and sometimes use intentional signs. In the first place, the will may simulate a passion; it feigns it by using the signs that testify quite naturally to that passion: "As for the laughter that sometimes accompanies Indignation, it is commonly artificial and feigned. But when it is natural, it seems to spring from Joy."[33] The fake laugh is the laugh of Democritus, which combines indignation and mockery because he bore ill will toward those he saw committing errors. The natural laugh is the laugh that seems to proceed from joy, for it can as well proceed from hatred, from admiration, aversion, or hunger. In any case, it is a symptom, since it relates to a cardiopathy. The laugh proceeding from joy has a physiological cause: the quantity of blood issuing from the heart fills up the lungs and the displaced air makes the muscles of the diaphragm and related facial muscles move. Given the difference in causes, Descartes rightly distinguishes the false laugh from the natural laugh, and even several sorts of natural laughs. But he was not in a position to differentiate them on the basis of their phenomenal appearance. The face presents signs that are marked by semiotic ambivalence. However, with the elucidation of the physiological mechanism of laughter, Duchenne could see this expression as the incontestable sign of joy, relative to which the artificial laugh could be easily identified. "It will be simple for me to show that there are some emotions that man cannot simulate or portray artificially on the face; the attentive observer is always able to recognize a false smile." The inertia of the inferior palpebral orbicular, which does not obey the will, unmasks the false laugh, that of Democritus.

But the will may also feign a passion by using the same signs that serve to declare it. For example, there are the wrinkles on the angry forehead of those who want to take vengeance in no other way than by looks and words. Descartes correctly integrates facial expression under the rubric of authentic signs, signs of a silent language serving to communicate. So it is not necessary for the expression to express an affective state for it to be signifying. To assume the mask of anger is to signify a passionate movement that one feels, or that one has never felt except as a feint. There is no semiotic ambivalence, for each person perceives clearly what the signs mean to say. But the difficulty lies in the issue that cannot be decided: do these signs arise from an authentic declaration or from feigning? Because the speechless language of deliberate expression proceeds from the will, it does not have the privilege of saying what is true. Descartes, as we have seen, was arguing against Cureau de La Chambre, for whom nature has made man's forehead and eyes speak in order to give him away when his words are not sincere. Duchenne does not argue that the will might allay suspicion by summoning to the face signs of passions to render them as if natural. Had he not registered the coincidence between natural expressions and simulated ones? What he was denying was the possibility that feigning annuls the primacy of the expressive function. From the moment the line of separation no longer passes *inside* the will, but passes between willed actions and involuntary expression, then the argument about a perverse will loses its meaning. That the simulated expression (mimicry) might be confused with authentic expression (natural language) does not obliterate the difference between them. Duchenne thus turns this indecision to the benefit of his physiognomic theory. The identity between the mask and the natural expression testifies to a specific, even inescapable, mode of expressivity. The fact that the latter is irreplaceable, immediate, and involuntary gives the signs of passion the character of biological events. Undoubtedly the field of expression may be employed by a will that reorients it to its desires. But this margin of maneuver does not cast doubt on the primacy of natural expressions. The will can mislead only by slipping into the forms already given by the expressive function. "The patterns of expression of the human face cannot be changed, whether one simulates them or actually produces them by an action of the soul. . . . This language must be always composed of the same signs, or in other terms, must be dependent on ever identical muscular contractions."[34]

Hence the third postulate that grounds a biology of the passions: in making the language of action depend on instinct, the Creator has also fixed the rules of its functioning. Analysis of the mechanism of human facial expression "demonstrates the principal facts of the grammar and orthography of human facial expression with the most complete empiricism."[35] What Duchenne discovered was not only the primacy of a natural language but the fact that we are already dominated by this language.

This interpretation was formed across the nineteenth century, and Friedrich Nietzsche showed that it would be wrong to see it as a form of spiritualism: "I fear that we will never get rid of God since we still believe in grammar."[36] If the sign does represent and if it has a content, this is because it is part of the grammatical organization by which this natural language assures its own coherence. For the sign to be able to say what it ought to say, it has to belong to a grammatical totality that is primordial, fundamental, and determining in relation to it. The language of passions is defined by the way in which it links all the elements that compose it to each other. Within an expression it is always possible to find the element or the ensemble of elements on the basis of which it was formed. But the material unity constituted by the play of muscular contractions is not governed by the simple combination of elements. It has its own principles: grammatical composition has regularities that are not transparent except to the psychologist. This orthographic image was transposed without any essential modification into the definition of psychological perception. It is a descriptive syntax of a language that has no history. The etymology and semantics of emotional language were not what they will be for Darwin: traceable back to the descent of man. For Duchenne it is only a matter of seeking the significations of an original language.

Since the eighteenth century, each physiognomic movement had been perceived as the image of a passion and each passion had its characteristic signs. Diderot, like Buffon, insisted on this evidence: "A man experiences anger, he's attentive, he's curious, he loves, he hates, he's condescending, he's disdainful, he's admiring; and each of these shifts in his soul paints itself on his face in terms that are clear, self-evident, about which we're never mistaken." But each expression, because it results from a movement of the soul, also designates the mechanism required to explain the diversity of expressive features. One thus moves from expressive movements to *the facial expression*. The play of passions ends up imprinting its stamp on the physiognomy. "Sometimes we set a physiognomy for ourselves. A face that's grown accustomed to the expression of a prevailing feeling will retain it." By force of animation, the face becomes a characteristic figure. On the one hand, Diderot was invoking Nature as giving a person a face corresponding to his natural penchants: "If a man's soul or nature has stamped his face with an expression of benevolence, justice and liberty . . . such a face is a letter of recommendation written in a language known to all men." But sometimes Nature gives a man a physiognomy that does not accord with his natural penchants: "Sometimes we receive it from nature and have no choice but to keep it as given. She decided to make us good but grant us wicked faces, or to make us wicked but grant us faces bespeaking goodness." On the other hand, Diderot was invoking History: sometimes the physiognomy is fashioned by the pressure of living conditions that act as factors that distort, or rather, transfigure: "In the heart of the Saint Marcel quarter, where I lived for a long time, I saw many children with charming faces. By the age of twelve or thirteen . . . they'd taken on the physiognomy of the market and in exchange . . . they'd acquired an air of sordid calculation, impudence, and wrath that would remain with them for the rest of their lives."[37] If the expressive face is marked by internal or acquired characteristics, it is impossible to *envisage* an expressive function. With the discovery of the mechanism of facial movements, Duchenne subjects the face to the sovereignty of function. In their fleeting expressivity, the play of features becomes preponderant. The character of a face is no longer what is dealt out by nature, at the risk of a misdeal, nor reformed by the style of life. For the essence of passions, whether given by nature or the effect of a metamorphosis, Duchenne substitutes the interplay of a function.

Again, we may measure the immense gap between Duchenne and the Classicists. If it was true that goodness could be masked by indices of wickedness, not only would Nature allay suspicion, but it would also annul the expressive function. This case is merely the naturalist version of the Cartesian doctrine of a perverse will. One immediately sees why Duchenne could only refute Diderot's assertion that Nature might betray us by inflicting on us a face that signifies the inverse of what we are feeling. It is true that each face carries a character, but that character results from the *habitual* play of feelings. What is given by nature is not a face in accord (or not) with our so-called natural penchants, but rather an expressive function. To describe the emergence of the face, it suffices to relate it to the expressive force. The child is not born good or bad, and man does not bring at birth his dominant passions along with the corresponding face (or not). "In the newborn, the soul is bereft of all emotion and the facial expression at rest is quite neutral; it expresses the complete absence of all emotions." Through usage, or non-usage, the function creates the physiognomy, the expressive face, the result of a *gymnastics* of passions with its imprint: "Facial expression is formed in repose in the individual face, which must be the image of our habitual sentiments, features of our dominant passions." The caprice of nature imagined by Diderot will never occur. Presupposing that in biology nothing is impossible, one would be confronted with an anomaly that is not serious. In admitting that nature has given us the face of a wicked person, this kind of figure would not be of the order of a morphological variation able to block the expressive function, not a normative variation. By contrast, the play of the expressive function, because it is normative, would have a corrective effect: "If a good man can be born with a wicked face, this monstrosity would sooner or later be effaced by the increased influence of a good soul."[38] With his old cobbler, Duchenne was not looking for an unpleasant male face, marked, moreover, by his social class, his occupation, and his age. In him, Duchenne saw the signs of a language common to all men, whatever their facial morphology, their social conditions, and their life history.

Physiognomic play designates a language of action governed by instinct and subject to the rules of its grammar. The final postulate was that expression escapes a mechanistic biology. The expressive apparatus translates a polarity, since it functions according to the norms of a harmonious language. The expressive function unfailingly satisfies the rules of a language. Muscular constrictions are fundamental elements, but they are insufficient to portray correctly a passion on the face. These elements have authenticity only when they give the corresponding signs their harmonic value. The language of passions is not only a text composed of characteristics; it has a vibratory nature that makes it akin to melody. Its expressive quality is linked to its tonality. Verneuil had stressed this essential characteristic of expression: "The grimace is to expression what noise is to music; in both cases the harmony is lacking."[39] It was because he happened to be fabricating grimaces and examining facial pathologies that Duchenne identified facial expressions that come under harmonics. In the study of complex expressions he sought to obtain a harmonious (meaning "natural") ensemble that would respond to the opposite affections. If the harmonious is the key to the natural, this is because the criterion of harmony is given by the structure of indication as it appears in normal life situations. Here expression implies the imbrication of a facial movement and an affect. The spontaneous play of the expressive function is the touchstone of correct expression. But Duchenne was not denying the possibility of correctly simulating the passions. It is revealing that he sometimes perceives the simulacrum of a passion as a harmonious expression, but this was because his model, Talrich, called on sentiments that he honestly rendered with most of his expressions. By soliciting fictive emotions, the signs of passions were rendered as if naturally. Duchenne thus found a theme that Horace had developed in his *Ars poetica*: if the actor wants to touch spectators through the emotions of the character he is playing, he must feel them himself. By the alliance between the will and the provoked sentiment, Duchenne thought he was holding the theatrical use of the expressive function as closely as possible to what it should be: expression properly speaking and not a language of mimicry functioning as backup to spoken language.

In his *Traité des passions de l'âme*, Descartes had placed the language of facial expressions under the control of the will: a deceptive mastery of the passions was always possible. But Descartes recognized in feigning, at least in its *representation*, a positive value. If not, he would not have evoked the passions that might be aroused in spectators: "When we read of unusual adventures in a book or see them represented on a stage, this sometimes excites Sadness in us, sometimes Joy or Love or Hatred, and in general all the Passions, according to the diversity of the objects offered to our imagination; but along with this we have the pleasure of feeling them excited in us, and this pleasure is an intellectual Joy, which can originate from Sadness as well as from any of the other Passions."[40] Diderot, for his part, had pushed as far as possible the analysis of the ways and means of imitation. Great actors observe, reflect on, and exert themselves to imitate the exterior signs of the passions. The muscular apparatus becomes as docile as the muscles of the fingers in an accomplished organist. The actor David Garrick's face, for an interval of a few seconds, could shift successively through just as many passions. But the imagination of the actor had to be turned toward the sublime. "What is truth on the stage, then? It is the conformity of action, speech, facial expression, voice, movements, and gestures to an ideal model."[41] Descartes and Diderot did not think that the actor could feel the least sentiment in playing his role. On the other hand, they did not doubt that the passions represented on the stage excited various emotions among spectators, and so they drew a line between the stage and the theater hall, between the actor as master of the game, and spectators who are moved. From the moment when the feeling that is really felt, or only experienced in an imaginary situation, is apprehended as the guarantor of the harmonious expression, Duchenne grants to the actor what Descartes and Diderot refused him. The success of the performance presupposes that his will relies on a movement of the passion he is feeling, if not really, then at least in an imaginary situation—lacking which, the expression will appear false: "It is true that certain people, actors above all, possess the art of marvelously feigning emotions that exist only on their face or lips. In creating an imaginary situation, they are able, thanks to a special aptitude, to call up these artificial emotions."[42]

We see how Duchenne effects a final break with the Classicists. For Descartes, the imagination prepared the way for the pleasures of the soul: with the emotions set apart, they are of a purely intellectual order. By the distinction between emotions interior to the soul and the passions of the soul, Descartes showed that the will is free: a person may always be in the theater of his or her passions. An emblematic and recurrent figure, Corneille's hero, animated by the passion of lucidity, may contemplate the combat of which he himself is the theater. For Diderot, the actor's imagination aims at an ideal model. After Duchenne, the imagination is no longer the relay that leads (as Descartes maintained) to disincarnated emotion or (as Diderot would have it) to the ideal incarnation. Imagination is at the service of imitation, since it allows the actor to invent a situation in which the production of indices offers a point of contact with what they are designating. Correctness of expression, in its phenomenal appearance, relates to the fact that the biological is both instrument and effect of a movement of passion obtained by the animation of oneself. If the will wants to reach its goal, in this case the faithful imitation of passions, it must rely upon a movement of the soul. In the actor, the imagination awakens the artificial emotions that are expressed by the play of facial features. The latter put into movement the imagination of spectators and unleash their emotions. In short, Duchenne's discovery belongs to the same register as the one formerly made by the Stoics: "The discovery of incorporeal events, meanings or effects, which are irreducible to profound bodies as they are to high Ideas. Everything that can happen, and everything that is said, happens and is said on the surface."[43] But to the extent that it includes a share of the imagination, the actor's playing is creation. Irreducible to the mechanical imitation of indices, the art of the actor calls for no commentary whatever. It is in the practice of the plastic arts that Duchenne saw the privileged field of application for his discoveries. To render the modeling of expressive figures in painting, as in sculpture—was that not one of the fundamental problems of the Beaux Arts since Le Brun? From the instant that the doctor of Boulogne enunciated the rules for tracing the expressive lines of the face in movement, the orthography of the face became the art of correctly portraying the passions.

4

Aesthetics

In 1862 Amédée Latour, editor of *L'Union médicale*, centered his review of *Mécanisme de la physionomie humaine* on the question of the rendering of passions:

In general, artistic masters have marvelously revealed the fundamental lines of expressions, but almost all of them have neglected or have not perceived secondary lines, for which M. Duchenne greatly reproaches them. He wants it to be known that these are not a simple ornament or a fantasy of Nature. He is committed to showing that they enrich the fundamental lines by furnishing certain important pieces of information. These rules, he says, cannot threaten artistic freedom or stifle the inspiration of a genius; they would pose no more of a hindrance than do the rules of perspective, for example. Nor should it be believed, he adds, that each expression is going to issue from a single mold, so to speak; the play of physiognomy cannot be so simple or so depressingly monotonous. We feel relieved by M. Duchenne's reflections. The uniform, the common, and the vulgar in the arts make us afraid. We want the artist to have total freedom and total spontaneity. As soon as the rheophore claims to do no more than indicate what M. Duchenne calls the *fundamental lines*, without imposing them—lines that the Greeks knew perfectly well without electro-physiology and of which they have transmitted to us immortal examples—we no longer see any disadvantage to M. Duchenne's making them the object of his demonstrations. It is even a blessing that physiology should be in accord with artistic sentiment. But that is not, believe me, the true objection that might be made to M. Duchenne. . . . In fact, the essays he attempted on three celebrated works of antiquity—*Arrotino, Laocoön,* and *Niobe* (whose orthographic faults he says he has corrected)—will appear a little brutal to lovers of the Ideal.

To touch such masterpieces! M. Courbet would jump up and down with joy, but M. Ingres and the whole School of Beaux Arts?! As for me (whose unprofessional opinion can be of no consequence whatsoever), I admire the patience, the zeal, the intelligence and sagacity that M. Duchenne has exhibited in this work. At first a stranger to photographic procedures, he went to the most skillful photographers to produce the expressive muscular expressions that are born willy-nilly from his expert device, the rheophore. Not content with these tests, M. Duchenne became a photographer himself, and although he modestly recognizes that he has not yet reached perfection, he has already produced a splendid album including 74 figures representing the most accentuated types of the feelings and passions expressed by the facial muscles. But I would say "overly accentuated"—and there lies the defect of photography. Art is more discreet, more veiled, more idealist. There are the frightening figures of truth, but of an ugly truth. Art should aspire to beauty, and inverting the poet's thought, we would say: *Nothing is true but the beautiful, and the beautiful alone is pleasing.*[1]

By inverting Nicolas Boileau's aphorism—"Nothing but the truth is beautiful, and truth alone is pleasing"—Latour lost sight of one of the principal problems of classical aesthetics. He could not understand the answer Duchenne was giving to the question once posed by Le Brun: What foundation can be given to the art of representing the passions? In wanting to find beauty on the path of truth, Le Brun was aiming at the ideal of classicism. The expression of the passions is subject to laws that imitation ought to recognize and apply. His schemas presented the signs of passions such as they ought to be represented. But Latour sees nothing intervening, not even the slow flow of a history that begins with seventeenth-century academicism and ends with the last lights of neoclassicism. An unconditional partisan of the ideal, the School of Beaux Arts is for him the privileged site where the tradition of great art endures. It is not by chance that he invokes Jean Auguste Dominique Ingres, defender of the most traditional aesthetic norms and one of the signatories of the *Protestation émanée des grandes artistes contre toute assimilation de la photographie à l'art* [Protest by Great Artists against Any Assimilation of Photography into Art] (1862). Latour could acknowledge only a limited bearing to Duchenne's demonstrations. By indicating the fundamental lines, the researcher was confirming what artists had already known for a long time: "Lines that the Greeks knew perfectly well without electro-physiology, and of which they have transmitted to us immortal examples." If the masterpieces of antiquity attained perfection, then laying hands on them is an iconoclastic gesture: "The tests he has attempted upon three celebrated works of antiquity—*Arrotino, Laocoön,* and *Niobe* (whose orthographic faults he says he has corrected)—will appear a little brutal to lovers of the Ideal." But there is more: Duchenne presents "the frightening figures of truth, but of an ugly truth." Hence this strange paradox: where Duchenne showed expressive figures that obey the laws of physiognomy in movement, Latour instead saw images of an unbearable ugliness. Duchenne responded to his critic that the true is a condition of the beautiful: "Nothing is beautiful without truth." With this lapidary phrase, he summed up the already ancient ideas of Roger de Piles: "Nothing is good, nothing pleases without the true, which is reason, which is equity, it is good sense and the basis of all perfections, it is the goal of the sciences; and all the arts that have imitation as their object are exercised only in order to instruct and divert men by a faithful representation of Nature."[2]

The art of representing the passions belongs to a tradition going back to theoreticians of the Renaissance. For Léon Battista Alberti, Giampaolo Lomazzo, and Leonardo da Vinci, solutions consisted in the acquisition of skill. But beginning with Le Brun, the rendering of passion was linked with a body of knowledge. By basing the tracing of facial expressions upon knowledge of how they are determined, he codified a segment of pictorial vocabulary and broke with empiricism. Le Brun rediscovered a requirement already enunciated by Franciscus Junius: the painter skilled in his art should reproduce the image of the very things that are missing from the human gaze. And so until the middle of the eighteenth century, what art lovers, artists, and theoreticians thought about imitating the passions was inscribed within the framework fixed by Le Brun. But in the decades from 1760 to 1780 there appeared in various fields the conditions that would make it possible to abandon his diagrams. In the field of pedagogy, there was a reform in teaching inaugurated by the Comte de Caylus at the Royal Academy of Painting and Sculpture. In academic painting, Jacques-Louis David tested the limits of the traditional mode of figurative expression. And in the realm of aesthetic thought, Johann Joachin Winckelmann and Gotthold Ephraim Lessing's historical and critical thinking about Greek art undermined the principles of academicism. But it was the first decades of the nineteenth century before Le Brun's doctrine became definitively outmoded, and the works of Duchenne rediscovered a positive and magisterial form for knowledge about expression. This is the irony of history: Duchenne gave a solution to the problem of representing the passions at a time when those artists who were engaged in modernity were turning their backs on history painting. It was the School of Beaux Arts that would gather and disseminate his teaching based on photographs.

At last it will be necessary to dispel a misunderstanding that traverses the history of artistic photography. Duchenne has been reproached for the artificiality of his images. But it has scarcely been noticed that he is one of the rare practitioners who did not fall into photography's trap of believing that one might actually seize the real, take it on the wing. The new medium, to which his contemporaries wanted to attribute the ability to restore the power of material reality, will run up against emotional expression. The price to pay for overcoming this difficulty is what makes Duchenne's images beautiful. He knew how to lend to photography, to expertly composed scenes and the complicated ritual that allowed them to be fixed, the power of invention and of desire.

The Conference of Le Brun

In 1648, a small group of artists gathered around Le Brun and Martin de Charmois placed itself under the protection of Chancellor Séguier and were endowed with royal letters of patent. The term "Academy" evoked previous artistic academies founded in Italy and the recent French Académie Française, a creation of Cardinal Richelieu. This institution was able to shake off the yoke of the guild, the "Mastery" of painters and sculptors. In 1654, after the defeat of the Fronde rebellion and the restoration of Mazarin's authority, Le Brun obtained a renewal of the statutes of 1648. A new stage was reached in 1663: by a decree from the Council of State, all artists holding royal patents had to join the new company, whose ties with monarchical power were thereby strengthened. In effect, the Royal Academy of Painting and Sculpture won out over the guild. It was a matter of making contemporary art the legitimate heir of Greco-Roman antiquity. But artists also had to be given rational principles and rules in order to guide their creative activity. To the existing practices of making copies after antique molds and of studying living models had just been added the study of anatomy and perspective. Finally, the academy would be a place for discussion. The 1663 regulations foresaw the establishment of meetings to discuss detailed observations of the arts of painting and sculpture. The definitive institution of regular lectures took place in 1667, taking up a suggestion from Jean-Baptiste Colbert himself. Visiting the academy, the minister expressed the desire to have a colloquium once a month at which one of the best paintings in the King's Cabinet would be explained by the professor in charge. Colbert insisted that the decisions taken after each lecture be recorded and that they become positive precepts for young artists. Lectures, debates, and resolutions would be committed to writing. André Félibien, the historiographer of buildings, who had the rank of honorary councilor, was chosen as editor. In the preface to the *Conférences de l'Académie Royale de Peinture et de Sculpture pendant l'année 1667* [Lectures at the Royal Academy of Painting and Sculpture for 1667], he insisted on their pedagogic role.

The following year, Le Brun gave his *Conférence sur l'expression générale et particulière* [Lecture on General and Particular Expressions]. Ever since, art historians have tried to find out from whom he borrowed his elements of physiology and his definitions of the passions. Early on, Roger de Piles indicated one path: Le Brun had exploited Descartes's *Traité des passions de l'âme*. Hence a first version (of this lineage) traversed the history of art: "[the treatise of] Descartes, who shared the profound interest of his age in the *perturbationes animae*, was largely responsible for the special psycho-physiological character of the theory of expression during the last decades of the seventeenth century among painter-theorists of the Academy."[3] A second version is marked by the integration into the story of Claude Nivelon, who had defended Le Brun against attacks from his contemporaries. Thus one learns that after his *conférence*, some listeners claimed that Le Brun was applying ideas from the illustrious Cureau de la Chambre. Starting from there, some art historians invoke a double filiation: Descartes *and* Cureau de La Chambre. They point out that the latter had written (well before Le Brun) two books that were widely distributed: *L'Art de connaître les hommes* [The Art of Knowing Men] (1659) and *Les caractères des passions* [The Characters of Passions] in four volumes (1640–1662). They stress that Cureau de La Chambre was doctor to Séguier, the foremost protector of Le Brun. Finally, they recall that Le Brun had designed the frontispiece of Cureau's book, *Discours sur les causes de débordement du Nil* [On the Causes of the Nile's Flooding] (1665). All of these factors argue in favor of an intellectual relationship between Le Brun and Cureau de la Chambre.

Quite recently, Jennifer Montagu has identified both of these sources for the *conférence*: "While one may justifiably suggest that Cureau's work may well have inspired Le Brun's interest in the subject of expression, the basis of this reasoning was supplied by the most up-to-date of thinkers, René Descartes. It was from Descartes that Le Brun took not only the physiological structure of his theory, but also the concept of the man-machine that enabled him to reduce the workings of the passions on the human body to such neatly predictable formulae."[4]

But these versions are historically false, for neither the *Traité des passions de l'âme* nor even a synthesis between Descartes and Cureau de la Chambre is the basis for Le Brun's theory, which offered a theory of expression that remained in the direct lineage of the philosophy of Séguier's doctor. Descartes attributed movement to animal spirits, while Cureau de La Chambre identified the movements of the soul with the principle of bodily actions. Yet these elements contradict each other: the former refers back to extension and so excludes the latter—that is to say, movements of the soul. Descartes distinguished between the cause of passions and their seat. The cause of passions—which integrates the mechanism into their definition—comes from physiological determinism: "[passions] are caused, maintained, and strengthened by some movement of the spirits." But when the passions are referred back to the soul, they are called the emotions of the soul, because "of all the sorts of thoughts that it might have, there are none whatever that agitate it and shake it so strongly as do these passions."[5] The passions, unlike the thoughts that designate the actions of the soul, are thoughts that are found indirectly in the soul. Since the latter is not the cause of the passions and of the signs that accompany them, therefore facial expressions do not express the soul's emotions. Descartes and Cureau de La Chambre did not attach the same semantic value to the notion of the sign. For Descartes, the signs of passions are twofold: sometimes the ever-ambiguous effects of a mechanism whose source lies in the heart and sometimes the expression of declarative or deceitful thought. Cureau de La Chambre saw these facial expressions as the effects of an expressive movement of the soul and as the pathognomic signs of passions. By defining passion as a movement of the soul that resides in the sensitive part, Le Brun was following Cureau de La Chambre. By its movements, the soul causes the agitation of spirits, which in turn determine the actions of parts of the face: "As it has been said that the soul has two appetites in the sensitive part, and that from these two appetites are born all the passions, so there are also two movements in the eyebrows that express all the movements of the passions."[6] Le Brun thought of signs according to both the epistemological model of Cureau de La Chambre's physiology and the semantic structure of his pathognomy. The sign is the effect of a physiological determinism that has its source in the soul. This is why the sign expresses a movement of the soul, which defines the passion.

To understand why Cureau de La Chambre's doctrine offered the only satisfactory solution, one must first resituate Le Brun's project at its point of origin. At the start, the models supplied by academic painting were said to clarify the problem of representing the passions. By borrowing the norms in use, the codification of the vocabulary of emotions seemed to have started along the surest path. Since the Renaissance, the great artists had painted images that imitated the effect of the passions; it sufficed to study their paintings. In his commentary on *Saint Michael Trampling the Dragon*, Le Brun stressed that this angel too much scorned the enemy that he had overcome to vanquish him, "which Raphael has marvelously well represented by a certain disdain that appears in his eyes and in his mouth. His eyes are half-opened, and the brows form two perfect arcs that are a mark of his tranquility, as is his mouth, whose lower lip slightly juts out under his upper one, which is also a mark of a contempt he has for his enemy."[7] But the technique of tracing expressions is enveloped in the painted image itself—and hence is inaccessible. One has to turn to the recommendations given by the most eminent painters. Solutions were limited: an effort either to register fleeting scenes in order to transcribe them onto the canvas (Leonardo da Vinci exhorted painters to sketch from life the postures, gestures, and diversity of facial expressions) or else to copy masterpieces so as to rediscover the savoir faire of the great painters. Nicolas Poussin, for his part, counseled the observation of the great works of antiquity so as to be imprinted with strong images. But these precepts about imitating nature, which constituted the basis of academic pedagogy, led to an impasse. Students possessed no formulae that might guide them in the art of representing the passions—hence a lacuna in the teaching of fine arts. There were indeed some rules for learning to draw a portrait: the oval for the head, and divisions by lines to mark unvarying key points. But students had no technique that allowed them to represent physiognomic movements. It is revealing that when Augustin Carrache and Giovanni

Francesco Barbieri Guerchin took an interest in the face, it was in the representation of its parts—the ear, in particular, so difficult to draw, was given special schemas and molds.

It was necessary to begin further back. For Le Brun, any passion is an expression, but any expression is not a passion: hence there was a distinction between a particular expression and expression in general. Until then, imitation of passions responded only to the requirements of expression in general. The latter designated the production of effects procuring an illusion of reality. The painter searched for a naive and natural resemblance of emotions to represent. He endeavored to present expressive faces without artifice and as they are in a natural state. Leonardo da Vinci saw this as the superiority of painting over poetry. A painting acts on the power of sight; the picture puts things instantly before one's eyes. By representing natural signs, the image ends up being identified with the object to which it gives an iconic form. The idea prevailed that painting, unlike poetry, makes its imitations with natural signs. Later on, Jean-Baptiste Abbé Du Bos would insist on this contrast: "The signs that painting uses to speak to us are not arbitrary and conventional signs, like the words that poetry uses. Painting uses natural signs."[8] Le Brun would overturn all this by showing that the signs of representation are nearer to words than to things. We know that he subscribed to the definition of general expression: "Expression, in my opinion, is a simple and natural image of the thing we wish to represent. . . . It is by this means that the different natures of bodies are distinguished, that figures seem to have movement, and everything which is imitated appears to be real."[9] But he did not subscribe to these ideas without reservations. If the definition of general expression applied literally to a particular expression, Le Brun would not have deployed so much effort to forming the concept of the normal tracing of passions. The classical theory of imitation, which underlies the definition of general expression, seemed unacceptable to him. On the one hand, naive resemblance carries the idea of crude imitation. On the other, natural resemblance imposes the idea of a copy after nature. A pedagogy founded on principles of general expression is doomed to failure. It remains at the level of empiricism. Moreover, since the form of any individual model is imperfect, it would never be possible to attain knowledge of the abstract model, the very type of perfection. The question was how to promote a reasoned practice of drawing passions and to constitute a pictorial language whose transparence assures total readability.

Le Brun was proposing a definition of the concept of particular expression that was symmetrical with, and the inverse of, the concept of general expression: to wit, a conventional and knowledgeable resemblance of the passions to be represented. First of all, he contrasts conventional resemblance with natural resemblance. The assertion that painting makes its imitations with natural signs presupposes that one confuses the perception of an expression with its image. This slippage of meaning, from perception of natural signs to that of the images that reproduce them, did not escape Le Brun. Because the painting is a figure of representation, it is governed by the same operations as language. The representative function of the image turns it into a sign. Imitation is invention of a pictorial language founded on the conventional character of graphic representation. Therefore Le Brun stressed the close similarity between pictorial language and spoken language. In the painted image had to be seen signs having the same function as words. The represented expressions took the place of "discourse and words: painting has no other language nor other characters than these sorts of expressions."[10]

But Le Brun also contrasts knowledgeable resemblance with naive resemblance. As long as pictorial language remained so close to the natural language of the passions, the signs represented therein owed their representative power only to their fidelity to natural signs. Le Brun gives a symmetrical and inverse value to Nature and to pictorial language, to natural signs and the images that represent them. Natural, the sign is nothing more than an element taken from things and constituted through knowledge. Because it is prescribed, one must take it as it is given in perception: as imperfect, equivocal, if not indecipherable. This is because the impassioned individual adds a series of organic events that muddle the essence of passions. To know the truth of passions, the artist must abstract the body and its defects. On the contrary, when one chooses a conventional sign, one might envisage it in such a way that it would be simple, easy to recall, and of perfect legibility. This means that Le Brun gives the principle of imitation a new *point of application*: not the natural signs of passions, but the passions in the purity of their essential manifestation. The arbitrary is opposed to the natural only when it comes to designating the manner in which signs have been established. The arbitrary is the framework of analysis and the combinatory space through which Nature gives itself just as it truly is—hence, the elaboration of a pictorial vocabulary founded on a knowledgeable resemblance. It is important to grasp not the appearance of expressive phenomena, but rather physiognomic phenomena such as they result from the causes that produce them. Art may attain beauty to the extent that it governs its creation by the laws of the production of essences that are revealed to it by physiological science.

The path that Le Brun was heading down was wholly traced in advance: the expression of passions ought to be governed by the principles that Cureau de La Chambre had fixed. Since the movements of the soul define the passions, it is through the diversity of these movements that they should be distinguished from each other. But it is the impression that one receives of the play of physiognomy that is the source of the analysis of the soul's movements. It suffices to consider the agitations that the body suffers within passions to discover the different movements of the soul, which are nothing more than the transcription of all that is most visible within the manifestations of the passions. The perception of bodily actions is the immediate apprehension of motor qualities. In the case of love, man is carried toward good; in hatred, he moves toward evil; in pleasure, he spreads himself on what is good; in pain, he avoids the evil that assails him. By the analysis of the causes that produce them, these manifest movements are first transferred from the exterior to the interior, from the domain of perception to that of explanation, from the visible effect to the invisible displacement of spirits. These latter are capable of four very simple movements that are common to all natural bodies: rising, falling, depleting, and condensing. Since it is the soul that communicates to spirits the agitation from which it suffers, its actions quite naturally present the same movements. From the center to the periphery: in hatred and in fear, the soul retreats or goes back into itself (falling); in pain and constancy, it contracts and closes in on itself (condensing). What were only impressions, by penetrating into the field of the invisible, become the dynamic of spirits. The actions of the soul are ultimately apprehended as images: "One might say that Love dilates them, that Desire springs them forward, Joy spreads them, Hope holds them firm, Bravery pushes them, and Anger brings them to the boil, and so on."[11] Thus is established a system of simultaneous presence that is, on the side of the effect, a perceived quality, and on the side of the cause, an invisible dynamic. One induces the causal image on the basis of the familiarities of perception. And one deduces the symptomological singularity of passions from the physical properties that one attributes to the image-movement. The system of causes is just the inverse of the empirical recognition of symptoms, a causal valorization of motor qualities.

Cureau de La Chambre had presented his descriptions of the passions as "tableaux" or "portraits." The latter showed the impassioned man in all his physical and moral states. The problem posed by Le Brun is more circumscribed: to give the formula for representing passions by relying on the soul's movements alone. The representation of schematic faces is only the graphic transcription of various movements identified by Cureau de La Chambre. The mechanism of how expressions are produced permitted giving rules for how to *present* them, in which drawings conform to the essences of the passions and to the image of natural signs. But the drawn expressions indicate expressive values founded on conventional rules. A physiognomic movement, which is the natural sign of a passion, cannot be rendered except in the form of a sign of this movement, in graphics, the sign of a natural sign. Each part of the face in turn may be affected by one of the four variables identified by Cureau de La Chambre. This means that the movement specific to each passion is a graphic rendering that conforms to the movement of the soul and of spirits. The eyebrows rise, lower, advance, or pucker; the nostrils lift, go down, contract, or dilate; the corners of the mouth draw down, up, withdraw, or move forward. Transposed into pictorial language, expression appears in the purity of its essence (Plate 7). Moreover, the combinatory of signs allows a modulation of the intensity of the expressions and an integration of the passions characteristic of history painting: ecstasy, faith, and veneration. It brings the whole field of the visible and the possible back to a system of variables, all of whose values may be transposed into pictorial language. The signs that represent the movements of each passion are all the more clear in that they are distributed along eight horizontal lines dividing the face into parts. For the profile, the vertical line passes from the external corner of the eye. These lines facilitate reading transformations in the face according to the movements of passions. They constitute the reference points indicating the position of traits, allowing one to follow the passage from one expression to another by the modification of the indicators. The reading of diagrams is governed by the same principles: a full face pivoted a quarter turn presents a profile, and coming back to the full face offers a nuance of the initial expression.

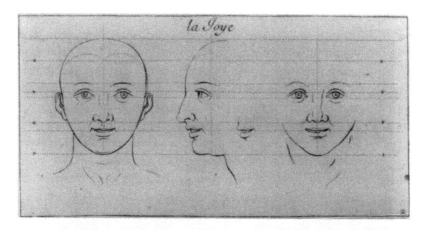

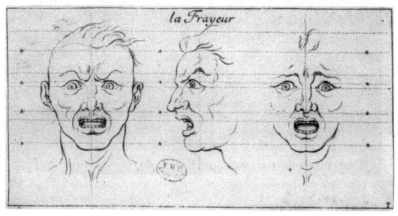

PLATE 7. Charles Le Brun, "la Joie" (Joy) and "la Frayeur" (Fright). Reproduced by permission of Bibliothèque de l'académie nationale de médecine/Université René Descartes, Paris.

It is true that Le Brun had already studied or copied the beautiful works of antiquity—Raphael, Pierro de Cortona, El Greco, and Poussin. The Battles of Alexander series would suffice to give a demonstration of the method of academic classicism. Some examples of the passions are directly drawn from *The Family of Darius before Alexander* (or *The Tent of Darius*) (1661). But these drawings of heads on black stone, by their manner, form an ensemble different from the one that illustrates the *Lecture*. In the mastery of the indicators of passions, Ernst Gombrich saw a good example of the application of a procedure from empirical experimentation: "The method used by Le Brun is all the more interesting in our context because it, too, is based on the study of art rather than on the observation of living expressions. Le Brun compiled a pattern-book of typical heads . . . and then proceeded to analyze these heads in order to find out what it was that made them expressive."[12] If the diagrams were the result of the application of a procedure indicated by Gombrich, one would have to take the *Lecture* as research—after the fact—with the imprimatur of science, in which case Le Brun's pedagogic project would remain founded on what he wanted to reject: an empirical approach closely linked to his interest in the noble style. We must resist the temptation of seeing the way Le Brun theorized his drawing practice as motifs of high rationality, serving only to protect his educational goals from any suspicion of empiricism. Hubert Damisch is also wrong to see his schemas "as 'false faces' on which are imprinted, as on theatrical masks, the graph of the soul's affections."[13] Assuredly, Le Brun remained a prisoner of seventeenth-century rules of seemliness and rhetoric. But a codification of expressive prototypes that chooses to rely on Cureau's physiology has quite another significance. Although this operation remains dominated by the intellectual and artistic culture of his time, nevertheless the concepts of truth and of beauty are freed from history. For Le Brun, they acquire a universal value. Because the purified forms of the passions are the graphic transcription of a passionate mechanism, they escape the double trap of empiricism and of servile copying. Le Brun's project presupposes the conjunction of pictorial practice and physiology. More fundamentally, it presupposes a search for the conditions that assure the alliance between the beautiful and the true. Le Brun's ideal was that of Nicolas Boileau-Despréaux.

Rendering the Passions

Very early on, Le Brun's pedagogic project encountered resistance. In opposition to the dogmatism that was leading national art in the direction of mannerism, André Fontaine offered a liberal alternative. In critiques from Henri Testelin, André Félibien, or Roger de Piles, the historians of art have thought they discerned a rejection of Le Brun's doctrine, a refusal of formalism, and the start of an emancipation from the rules that he had fixed. Quite recently, Édouard Pommier insisted again

on this point: "The great theorists of the classical period, like André Félibien and Roger de Piles, are too intelligent, too sensible, and already
too 'historically-minded' to not have reservations about these attempts at
regulatory power that would turn the painter into an interpreter of a pre-
established normalization."[14] Later on, the codification of passions in the
manner of Le Brun would provoke a fully commonsensical reaction. Artists were supposed to turn toward the oldest and surest solutions: copy
the paintings of the masters, draw the living model, or make their expressive self-portraits in front of a mirror. But art historians also agree in recognizing that the drawings of Le Brun enjoyed an unprecedented vogue.
In the last half of the seventeenth century and throughout the eighteenth,
his figures would be unavoidable. His influence is measured by the quantity of related publications, in both France and abroad. Various editions
of the *Lecture*, with engravings of his drawings, were published by Testelin (1696), Le Clerc (1696), Picard (1698), and Audran (1727). Numerous
re-editions and a series of translations into English, German, and Italian
assured their widespread distribution throughout Europe. Identifying his
schemas in the paintings by several artists would also establish his influence. We may cite *Alexandre et la famille de Darius* (1680) by Jean Jouvenet
and the *Bacchus et Ariane* (1693) by Antoine Coypel. In the eighteenth
century, there is Jacques-Louis David's *Le combat de Minerve contre Mars*
(1771) and *Erasistrate découvre la raison de la maladie d'Antiochius* (1774),
and Joseph Benoît Suvée's *L'Amiral Coligny défiant ses assassins* (1787). We
recognize Le Brun's formula as well in paintings by Jean-Baptiste Greuze,
La malédiction paternelle (1777), by Jean-Honoré Fragonard, *Au génie de
Franklin* (1778), and by Joshua Reynolds, *Mrs. Siddons as the Tragic Muse*
(1783). Hence the paradox underscored by Montagu: "Throughout the
eighteenth century one finds the strange contradiction of students being
trained in Le Brun's illustrations, and many artists manifestly imitating them, while the same illustrations were generally condemned, with
increasing force as the century progressed."[15]

However, this historical version is not credible, and the paradox is
merely apparent. First of all, historians have not understood that Le Brun's
aim was not to furnish a model for the *painting* of passions but to provide
a solution for the problem of representing them. Le Brun knew very well
that his schemas might be perceived as abstract, stereotyped, and disincar-
nate. But it was by design that he offered them in this form: patterns on

the basis of which the finishing of the painting might give it more flesh and more life. As for the alternative solutions that were proposed, it was precisely those from which Le Brun wanted to free students. On the one hand, study of the masterpieces of the Renaissance might contribute to forming artists' tastes, but it would not teach them how to *make* expressions. On the other, the technique of the self-portrait proposed by Roger de Piles and later by Gérard de Lairesse did not escape empiricism. And then, trying to measure Le Brun's influence by the number of editions of his book in France and abroad is a hazardous project. Nothing about how a book was used can be deduced from details about its dissemination. Was it consulted by artists or by art lovers? It could be supposed that its place was in the libraries of the curious. Finally, one would be wrong to seek the traces of Le Brun's influence in the paintings of his successors. Except for the obvious application of his schemas by some painters like Louis Simoneau, Suvée, and David in his early period, the iconographic analysis is uncertain. On this point, the divergence of opinion among the Classicists about the compositions of Greuze is already revealing. For example, the engraver Pierre Charles Ingouf drew his expressive heads from Le Brun, and Diderot invented his own pictorial vocabulary by drawing from nature. The register of passions is too limited and too constraining for one to conclude with any certainty that there was generalized use of his schemas. "Facial airs" necessarily have family airs. One should not ascribe the uniformity of the results to the application of a standard model inherited from Le Brun. It is useless to insist on the presuppositions of such a historical method. It suffices here to show its consequences: the need to record as observed facts the conflicts between the partisans of Le Brun and his opponents; the difficulty of grasping the reasons for the oppositions between those who accept his diagrams and those who continue to exploit the more traditional solutions. Finally, one is obliged to leave in suspense the question of when and how Le Brun's drawings were cast aside.

In the eighteenth century, there were not two currents of thought that confronted each other over the subject of representing the passions, but rather two different ways of envisaging them. Le Brun's preoccupation was to give examples to facilitate apprenticeship in learning to trace the passions. But what he had not foreseen—and something that art historians have not seen—is that his illustrations might be appreciated on the basis of inappropriate criteria drawn from academic painting. If you search his drawings for an ensemble of the elements that are found only in a completed painting, you are bound to be dissatisfied. The reactions of Testelin, Félibien, and de Piles, far from opening a new era, are the signs of a basic misunderstanding. That it was necessary to take into account the characters of figures, proportions, and temperaments, as Testelin said, nobody doubted. For their part, Félibien and Roger de Piles were content to reactivate the ancient precepts: the former said that images of passions might be rendered by the imitation of nature, and the latter that painters might make beautiful expression by taking their imagination as a guide. From a practical point of view, these remarks were of no utility—hence the idea that one had to study the Renaissance masters to better understand the diversity, correction, and elegance of expressions. It was toward Raphael that Mademoiselle Le Hay (Elisabeth-Sophie Cheron) turned in order to illustrate her *Livre à dessiner, composé de têtes tirées des plus beaux ouvrages de Raphaël* [Sketching Book, Composed of Heads Drawn from the Most Beautiful Works by Raphael] (1706). Later Benjamin Ralph published a more complete selection in his *The School of Raphael: Or the Student's Guide to Expression in Historical Painting* (1759). On the other hand, many were those who saw in Le Brun's illustrations an unprecedented solution. For a theorist like Du Bos, talent consists of giving life to people through expression. The painter should apply to the head that he is doing what Le Brun says about the effect of passions on the face and the traits that mark them. William Hogarth, who was interested in the linear description of the language of passions, found a good point of departure in Le Brun's drawings. What seduces Diderot in his schemas is the correct relation between expressive movements and history. In the article "Passions" in the *Encyclopédie*, Claude Henri Watelet follows Le Brun's lesson and Chevalier Louis de Jaucourt follows Watelet. Around 1780, Camper would write: "Nobody has treated this subject with more order than Le Brun in the middle of the seventeenth century, and consequently more than a century ago. One might add to his glory that all the Nations have taken as the basis for their teaching not only his precepts but even his drawings, which have become a sort of universal type."[16]

At the time when Camper was still celebrating Le Brun's merits, some events had already occurred that would make it possible to abandon his schemas. These events are known, but they have to be put into contextual relation in order to perceive the convergence of their effects, both innovative and destructive: a new way of representing the passions and the ruination of Le Brun's doctrine. In the realm of teaching, Caylus reintroduced the study of expressions from models. At the start, his preoccupation was no different from Le Brun's: perfecting the noble genre of history painting. Later, the new requirements of the Expressive Head competition would end up making Le Brun's images unusable. In academic painting, David could copy from antiquity the calm and dignified qualities that one sees on his characters' faces. He also knew how to give them expression—in the manner of Le Brun. Later, his reaction against academicism would lead him to a radical conversion, and so during his period of exile, he painted from a model: one common model with no particular quality. Finally, the critical analyses of Winckelmann and Lessing struck a decisive blow against history painting. But in stressing beauty to the detriment of expression, they would provoke an unexpected reaction. Goethe and Charles Bell would relate the emotions of the soul to their bodily manifestations and would reestablish the link between art and knowledge. It would not be unfair to say that these three transformations appeared within the framework of neoclassicism. It is not by chance that there existed common preoccupations among amateurs, painters, and scholars. Caylus shared Winckelmann's interest in the study of antiquity. David participated in the competition of expressive heads instituted by Caylus and, during his stay in Rome, he became aware of the theories of German thinkers. But these relations should not mask the specificity of the various avenues and means that led to the rejection of Le Brun's schemas.

To describe the first factor, education, one must start with a modest pedagogic innovation. In 1759, the Comte de Caylus established the "Concours de la Tête d'expression." He was pursuing the same goal as Le Brun: to advance instruction in the arts by making it easier for students to study heads and expressions. But Caylus reproached Le Brun for having given ideal images that were distant from Nature: "Le Brun felt the need for such a study, he wanted to make good this lack by the traits of passions and the heroic characters that he had had engraved. It was of mediocre help, and you know, gentlemen, what utility these traits might have; when they are not so strongly subject to mannerism, what are they in comparison with nature?" So there was a return to one of the habitual prescriptions of imitation: the faithful reproduction of the model. The academy opened up to students a competition to draw or paint, to sculpt in the round or in bas-relief, a head from Nature, life-sized and representing the expression of a passion. For Caylus, expressions like contempt, disdain, and indignation were easy to imitate. On the other hand, as Alberti had already noted, expressions of pain and pleasure are often so difficult to distinguish from each other that art could not undertake to render them without accessories. The readability of these expressions would be facilitated by decorum. From a pedagogic standpoint, there was another advantage: the professor would explain with reference to each figure the reasons for the nobility and beauty of the face. He would also take care to pose the different expressions by always relating them to a known subject taken from fable or from history. As a model, one should choose a young man or young woman with an advantageous physique, who would assume a pose: "He [or she] should appear naturally unadorned and with hair in ringlets or arranged picturesquely, or should be coiffed with ornaments suitable for heroic drawing, that is to say, a Greek helmet for Achilles' anger, for Pallas Athena's pride, or for Mars' ardor in combat, etc. One would also pose the character of a bacchante coiffed with ivy or vines, the head thrown back to express as much as possible to someone of sangfroid the species of enthusiasm felt by the followers of Bacchus, who were not simply women seduced by wine."[17]

But the study of passions in the studio ran up against difficulties. When an expression is drawn from a life model, that person is immobile. Should the imagination supply what is most difficult in the expression? In that case, the exercise would be fruitless. Or else should a model be brought to the studio who acts out the passion? Even if the model is a good actor, the expression quickly turns into a rictus. Cochin remarked that the results were not up to the ambitious reform. Experience showed that the competition had not provided the advantages that Caylus expected from it. Despite these criticisms, the form of the competition would long remain unchanged. But its subjects would change. To take the facts in strictly chronological order: from 1760 to 1791, when the competition was suspended (followed two years later by its suppression after the decree from the revolutionary Convention of August 8, 1795), the themes proposed were the simple passions with which Le Brun had been concerned: Gaiety, Attention, Contempt, Fear, Astonishment, and Admiration. In 1796, the minister of the interior Pierre Benezech authorized professors to resume the competition founded by Caylus, using still the simple passions taken from Le Brun's register, but also variations like "Surprise mingled with joy,"

"Sleep diverted by an agreeable angel," "Heavenly contemplation," "Attention mingled with fear." After 1813, more time was given to the competitors to prepare their entries, and one of the modes of competition that its founder had envisaged was finally adopted. Caylus had wished that before starting, the professor should address the students by reading aloud to them the author and passage from which he had taken the situation as context for the chosen expression. By attaching an emotional state to a story, the students were expected to have greater finesse in rendering expressions. The themes of the competition grew more precise: "Psyche, examining Cupid, is seized with admiration at the sight of an object that she had thought a monster" (1813); "The Virgin Mary at the moment the Angel Gabriel announces that she will be the mother of God" (1821); "The Pain mingled with Joy of a Martyr" (1823); "Rage mixed with Contempt: Philoctetus before Ulysses" (1829); "Faith mingled with Hope" (1830). Therefore, the jury wished students to represent a complicated and subtle psychological moment: not the expression of a passion, but an emotional state that designated the individual portrait of a passion. It was in the form of a piece of history painting that the jury was called upon to judge the rendering of a model incarnating a passion. Let us recall the essential point: the choice of subjects for the competition reveals a transformation: the substitution of a rendering of a *complex* of emotions for the rendering of a simple passion.

We turn to academic painting to identify the second event. David was trained within academic teaching and won the Caylus Prize in 1773 with a drawing of "Pain" (Plate 8). His first canvases show that he exploited all the resources of history painting. During the transitional period that corresponds with his stay in Rome, the use of academic procedures is still evident. In *Saint Roch Interceding with the Virgin to Cure the Plague-Stricken* (1780), one head translates extreme despair and another expresses astonishment with fright. Having pushed as far as possible the conventions of academicism, David was one of the rare painters to become aware of the cultural impasse to which they were leading: "People love theatrical effects and when one is not painting violent passions, when one is not pushing *expression* in painting to the point of a *grimace*, one risks being neither understood nor appreciated."[18] *Cupid and Psyche* (1817), despite its success when it was presented in Paris the following year, was really neither understood nor appreciated; it truly is a strange picture (Plate 9). Psyche is represented in a lascivious position; her left arm brought back to her head, she sleeps on. Cupid does not have the spiritualized grace that one expects to find in an

PLATE 8. Jacques-Louis David, "Pain" (1773). Reproduced by permission of École nationale supérieure des beaux-arts, Paris.

PLATE 9. Jacques-Louis David (French, 1748–1825), *Cupid and Psyche* (1817). Oil on canvas, 184.2 cm × 241.6 cm. Copyright The Cleveland Museum of Art, Leonard C. Hanna, Jr., Fund, 1962.37. Reproduced by permission of Cleveland Museum of Art.

adolescent who incarnates beauty and desire. His facial traits appear crude, his physique is immature and his complexion sickly. Cupid is neither a god nor a hero but a common model copied in servile fashion. The contrast is striking between the idealized beauty of Psyche and the realism of the god of Love. Evidently the norms of representing the myth are being transgressed: Psyche would indeed be right to think that Cupid is a bestial and repulsive figure. Still more unexpected is how the choice of the fertile moment, which held no secrets for David, here becomes the most preposterous moment imaginable. Cupid's posture is unsettling: his right foot already touches the ground but his leg and his left foot are entangled in the drapery on which Cupid is lying, also coiled under the left armpit of his lover. David has painted a moment that must turn, if not into a nightmare, then at least into ridicule. Cupid in leaving Psyche can only awaken her. This fertile moment is the instant before the grand crisis of a subject—a fatal crisis, since the painting can only disintegrate by running off with the allegorical genre. Of this there can be no doubt, given the clues David has laid. The only window in the room is a painting within a painting, a landscape that echoes the scene: dawn breaks and the sky is divided by two extinct volcanoes, the woman asleep and the man satiated. Psyche's soul takes the form of a vulgar butterfly, already pinned onto the bed by the fickle god.

David's contemporaries were deeply troubled by the figure of the god of Love. Painters and art critics probed this figure: it has a rather "faunesque" character (Gros), it offers a "mocking expression" (Kératry), even "a cynical grimace" (Miel). This "imperfection" was corrected in Alexandre Giboy's engraving, published by Miel, in which the figure of Love has been transformed: the grimace has been replaced by an innocent and gentle smile; the torso and shoulder appear more developed, more muscular. Two years after the exhibition in Paris, the engraver Jean-Louis Potrelle also corrects the figure of Cupid, which provoked David's understandable irritation: these criticisms and corrections are evidence of incomprehension. People did not understand that *Cupid and Psyche* marks the death of the allegorical mode: "Our ideas are getting better, and often our latest works feel those effects. It is perhaps to this that I owe the success of my last work, happy if the execution, which does not always match the conception, manages to obliterate the age for which these sorts of works were conceived."[19] The strength of the painting resides in what the critics took for faults in the composition: there is no cynical grimace or mocking expression, but instead a contemptuous smile. The break with Le Brun is consummated. Not only did the latter describe the laugh and contempt separately, but these expressions are also marked by antagonistic movements. Indeed, the opened mouth expresses the movement of laughing, the corners pulled back and rising up—but the eyebrows are lowered on the side of the nose and the eyes almost shut. But in contempt, the mouth is closed and the corners slightly lowered, the lower lip exceeding the upper one. On the basis of Le Brun's schemas, the representation of a contemptuous smile is impossible. Yet by scrupulous imitation of the model whom David had asked to smile, the contemptuous smile is well rendered. It is the sort of smile where one shows one's teeth, without the eyes participating in the expression since they are opened wide. The perplexity of the critics is not surprising. Cupid's smile appeared to them to be *misplaced*: in the picture it marks his contempt for Psyche at whom he does not even look—the height of contempt. But Cupid's contemptuous smile is also unsettling; perhaps there is contempt for the spectator who tries in vain to find what David did not want to show: an amorous scene, erotic in the manner of Canova or Girard. Cupid's expression also sends the spectator back to his own image: the contemptuous smile that marks the superiority of man over woman, the sexual object. Finally, Cupid's contemptuous smile is heavy with *thought*: the moral condemnation of allegory by David the philosopher, who is adopting the posture of Democritus.

The third event occurred in the context of new aesthetic thinking. For Cureau de La Chambre, the expression of passions had its source in the movement that emotional tempest excites in the soul. He compared the latter to a great abyss that, without leaving its confines, suffers all the movements that this tempest can excite in there. A single metaphor suffices to undermine the foundations of the physiological theory on which Le Brun was leaning. Winckelmann orders his perception of the relation between depth and surface around a requirement of an ethical order: "Just as the depths of the sea always remain calm however much the surface may rage, so does the expression of the figures of the Greeks reveal a noble simplicity and sedate grandeur even in the midst of passion."[20] Hence the opposition between the wise discretion of the ancient artists and the works of the moderns, in which one perceives overly marked traits—an extravagance that borders on deformity. The figures of expression resemble the masks of stage actors obliged to make themselves comprehensible to spectators in the farthest rows. That is not all: "This exaggerated expression is even taught in a book that is put into the hands of beginning students of art, namely Charles Le Brun's treatise on the passions."[21] The more that expressive movements are manifest, the more they alter the harmony of forms and move away from that noble simplicity that should be the goal of art. The bodily manifestations of the passions ought to disappear in favor of the representation of an inner state that marks the mastery of emotions. Moral and physical pain, even a devastating one, should be expressed by an impassiveness that touches on insensibility. Winckelmann, in his *Réflexions sur l'imitation des oeuvres grecques en peinture et en sculpture* [Reflections on the Imitation of Greek Works in Painting and Sculpture] (1755) remarked that if statues of *Laocoön* sigh but do not cry, as Virgil has him do, it is because the artists want him to express the soul's strength and grandeur. *Laocoön* is ideally beautiful not because it is the image of suffering but because it illustrates the lessons of stoicism (Plate 10). Later on, Winckelmann went further in his analysis of *Laocoön*: "The chest strains upward with stifled breath and suppressed waves of feeling, so that the pain is contained and locked within."[22]

But Lessing notes that the fact of crying in physical suffering is not incompatible with grandeur of soul. The artist should thus have another reason to abstain from making his marble figure cry. To represent the beauty most compatible with physical suffering, one had to moderate the strength with the suffering and produce a sigh rather than a cry, "not because screaming betrays an ignoble soul, but because it distorts the features in a disgusting manner. Simply imagine Laocoön's mouth forced wide open, and then judge. . . . A wide open mouth . . . becomes in painting a mere spot and in sculpture a cavity, with most repulsive effect."[23] Essentially, Lessing subjected expression to the demands of the key moment: art could only give an immutable duration to what is transitory. Not only might the cry be perceived as a sign of weakness, but it would fix the imagination on the coming death. The representation of the natural language of the passions is incompatible with beauty. For this reason a *Laocoön* screaming with pain, as well as the portrait of Julien Offray de La Mettrie laughing as Democritus, cannot truly become masterpieces. But Winckelmann and Lessing, concerned to put an end to a rhetoric of bodily events, had neglected an aesthetic principle to which Goethe would soon attract attention: "The greatest works of art show us highly developed human forms. We expect

PLATE 10. Guillaume Duchenne, *Le Laocoön*. Reproduced by permission of École nationale supérieure des beaux-arts, Paris.

above all a knowledge of the human body, of its parts, proportions, its internal and external functions, as well as its forms and movements in general."[24] Only a physiologist would be able to describe the intimate union of soul and body, and thus show that effort and suffering are subject to the exigencies of a physiology of emotions. Not by chance: Charles Bell shows that if the *Laocoön* does not cry out, it is because the cry is here a physiological impossibility. The artist gives the impression to spectators that the *Laocoön* suffers in silence and that his expression is not exaggerated. This is because he blocks his respiration to give more strength to the muscles of his arms, and the result is that no breath escapes his lips except for a faint moan. "It is a mistaken notion to suppose that the suppressed voice, and this consent of the features with the exertion of the frame, proceed from an effort of the mind to sustain his pain in dignified silence: for this condition of the arms, chest, and face are necessary parts of one action."[25]

A new approach to expression: to represent the passion means to render a psychological and physiological complexity. Only copying from a model seems to offer new expressive possibilities. It is the individual who is represented in the Expressive Head competition. It is the individual who becomes the subject of David's paintings. Again it is the individual who is apprehended as an organism whose movements are regulated by the laws of muscular physiology. Goethe insisted on this ultimate requirement, without which the faithful imitation of a model could not reach its goal. To be more precise, one would have to say that he set the program around an aesthetic open to anatomical and physiological knowledge: "This exteriority, this envelope, is adjusted with such precision to the internal construction—variegated, complicated and delicate as may be—that it becomes itself internal, for the two determinations (interior and exterior) are always in direct relation, whether in the state of complete rest or the most violent movement."[26] Charles Bell is no doubt the person who pushed farthest the analysis of the connections between the soul's emotions and bodily signs. He could describe the physiological mechanism of a sigh, but he was unable to explain the physiognomic movements that carried so many essential meanings. The reason for this failure is of an epistemological kind: the lack of a descriptive anatomy of the facial muscles, since the interface between the muscles and the skin was still an enigma. Bell's neuro-muscular physiology was not up to demonstrating physiognomies for the use of artists concerned with explaining how the passions moved. The prestige of Lavater was the counterpart of this failure. As long as people were ignorant of the rules of how fleeting expressions were composed, then character alone was supposed to give the permanent image. Character was perceived as the result of the action of the dominant passions. We have seen how Moreau and Cruveilhier turned toward physiognomical theory, but they could do no better than the anatomist Jean Joseph Sue, who in 1788 recommended that artists consult the "celebrated Lavater." We have seen how Duchenne realized Johann Wolfang von Goethe's program by describing the adjustment between interior and exterior determinations. Now we shall see how this doctor of Boulogne wished to intervene in the domain of the plastic arts.

Expressive Heads

In his preparatory report for the decree of November 13, 1863, on the organization of the Imperial and Special School of Beaux Arts, Count Alfred Emilien de Nieuwerkerke proposed inviting some scholars able to expound new theories. Eclecticism seemed the order of the day: "We see little disadvantage in people developing very different systems within the same precinct, so that, for example, servile imitation of Nature and the search for an Ideal type are being preached in turn." At first sight, Duchenne's pedagogic project ought to have slipped easily into the instruction given to students. Moreover, the count was thinking of him. While belonging to the academy tradition, Duchenne was renewing the ways of responding to some of its purposes. Two centuries after Le Brun, Duchenne restated the question: what does one have to know in order to represent the passions correctly? By the elucidation of the rules of tracing expressive lines, he would be offering students the choice tool. They would be equipped to tackle the Expressive Head competition. In addition, students' interest in the history of art would be enriched by a critical look at masterpieces of the past. But after the protests that followed proclamation of the decree, Duchenne was not able to present his research at the School of Beaux Arts, where the reigning conservatism was scarcely favorable to the reception of the work of a doctor who claimed to be giving lessons to artists. In order for his discoveries and his photographs to become teaching material, there would have to be a reform. In a decision by the government of the republic in 1870, the School of Beaux Arts was again attached to the Ministry of Public Education. In 1872, Mathias Duval introduced Duchenne's work to the school: "There is no place more worthy of it, for it serves each year in teaching and, in perpetuating among artists the memory of the person who has finally based the study of expression on rigorously experimental data, it personifies for them the fertile union of Science and Art."[27]

There is a very general institutional reason to explain the delay in the acceptance of Duchenne's theory within the School of Beaux Arts, one that is somewhat external. A variety of particular causes may be considered. Curiously, the manner in which Duchenne solved the problem of tracing the passions placed him at the center of contemporary debates concerning the relations between anatomy and movement, between technique and art, the true and the beautiful. Duchenne returned the study of anatomy to primacy, insisting on the didactic value of his photographs, and he presented his grammar of the language of passions using the traits of a model of unpleasant physique. Thus the object of study, the technical procedure, and the subject of the experiment all gave grist to critics who were rejecting anatomy, photography, and realism. Duchenne placed anatomy at the heart of his theory of expression. In the eyes of those who remained attached to the artistic ideal inspired by antiquity, a pedagogic project founded on anatomical accuracy was unacceptable. Charles Blanc reports a significant anecdote: "One day Ingres entered his studio and saw some of his students off to the side drawing a reduced plaster cast of Houdon's anatomical model, and advancing toward them, he broke the plaster figure."[28] Duchenne was introducing photography into the plastic arts. For those who placed the powers of the imagination above the copying of details, this amounted to an attack on the dignity of creation by aligning art with a mechanical procedure. Dechambre stressed the preponderant role of the creative faculties: "It is sometimes necessary for the expression to augment the value of certain traits at the expense of others, and to create in this way a sort of artificial harmony that responds better, if not to physical reality, at least to the no less true model created by the imagination; one understands that it would be unreasonable to constrain art rigorously to a sort of tracing or embossing of the human figure."[29] Finally, Duchenne had chosen as model an old man whose ugliness thwarted the aspirations of academicism. In 1862, it was the Ingresque charm of the pictures of Pierre Puvis de Chavannes, Alexandre Cabanel, and Paul Baudry that was seducing the public. But a contrary movement grouped realist painters around their master, Gustave Courbet. At the moment when figures of idealized beauty were triumphing in the official salon, Duchenne's old model appeared vulgar and trivial: "People reproach M. Duchenne for having stripped art of any ideal and reducing it to an anatomical realism quite in line with the tendencies of a certain modern school."[30] But one has to look more closely; on these issues, the position of the doctor from Boulogne is more nuanced than his opponents are led to believe.

In the practice of the plastic arts, Duchenne did not believe any more than did Ingres in the utility of anatomical knowledge. One had to stick to study of the living model and observation of the person in movement. It is a long way from embossing to the human form, from the cadaver to the living person, and from the study of muscles to that of external appearance. In the exaggeration of anatomical science, Duchenne saw one of the principal causes of the decadence in art. He agreed with Diderot, who had already remarked that "profound study of anatomy has spoiled artists more than it has improved them." But he added that "in painting as in morals, it is quite dangerous to look under the skin."[31] Diderot's mistake lay not in distrusting depth but in not seeing that anatomy might be the study of *superficial* structures. With the examination of the *sublime* muscles—the adjective was precisely used in anatomy to mean "superficial"—Duchenne linked knowledge of muscular actions with knowledge of how expressions were produced. Therefore it is not surprising that he insisted on the symmetrical and inverse danger to what Diderot was talking about. It is dangerous to see only the skin—if not in morals, then in the plastic arts. In drawing the expressive lines of the face, many artists were losing their way. The anatomy of the facial muscles has the particularity that depth is only the inverse of the surface, in which the functioning of superficial muscles adds something to their movement: the expression with contractions. But the adjective "sublime" also designates what is the highest on the scale of aesthetic values: attitudes or "airs of the head." By the relation he established between muscular actions and facial expressions, Duchenne naturally placed himself on the terrain of elevated subjects: representation of the visual perception of passion. As soon as he announced rules for tracing the lines of a face in movement, the orthography of physiognomy became the art of correctly painting expressive lines. But Duchenne willingly excused the artist from having a profound knowledge of the anatomy of facial muscles, provided that he applied the rules that flowed from the relevant research. The drawing teacher should be an expert commentator on photographs—hence Duchenne's effort to reframe and reformat his photos. It was useful in teaching to enlarge heads that might serve as demonstration of the fundamental principles of his physiognomic theory. Photographic prints were becoming centerpieces of drawing classes.

The photographic document fulfills its function by teaching the student to see Nature. But does one therefore have to speak of the subjugation of art? Baudelaire said, "The exclusive taste for the True (so noble a thing when it is limited to its proper applications) oppresses and stifles the taste for the Beautiful."[32] Duchenne knew very well that the exact reproduction of nature was not art's goal. But he saw no reason why the artist should be deprived of a procedure that might serve the taste for the Beautiful. Quite the contrary—his photographic studies put under artists' eyes what escaped their observation. The finest perception of an expression could never capture the real changes in the face. The expression of a passion is too brief for an analytic view to be possible. Only the sensitive plate collects the image, fixes it, and renders all its truth. Moreover, the use of photography does not annihilate the play of creative faculties. The artist's imagination and freedom are not impaired. In the rules of physiognomy in movement, Duchenne saw something of the same order as the laws of perspective: knowledge about vision and a science of the graphic representation of space. Why could not history painting integrate his grammar of the language of passions, as it had integrated the laws of perspective? If science was being applied to the Beaux Arts, it was in order to serve them: "The rules of the mechanism of physiognomy, deduced from electro-muscular experiments, enlightening the artist without shackling the freedom of his genius." Duchenne's physiological thinking did not try to deny diversity as such, nor to challenge it, but to understand and provide grounding for it. The formulae of the expressive function, in their general form, were merely rules allowing one to determine which muscular contractions were harmoniously combined. By applying these formulae you could avoid errors. "Evidently, these rules cannot take the place of genius, but in teaching the art of *correctly* painting the movements of human facial expression and by making the natural harmony of its expressive lines known, these rules can prevent or modify the errors of the imagination." Finally, the strict observation of rules deduced from the mechanism of physiognomy is not opposed to the individual representations of the passions. The artist might vary the traits of the same passion: " Each expression does not come, as one might once have thought, from a single mold; the play of facial expression cannot be either as simple as that or as dreadfully monotonous."[33]

The photos presented a model with unpleasant physique. In stressing the ugliness of this model, Latour did not see that Duchenne was dissociating expressive beauty from formal beauty. In truth, nothing prevents the ideal of expression from being evident in a subject of coarse aspect: "I simply wanted to prove that, despite defects of shape and lack of plastic beauty, every human face can become spiritually beautiful through the accurate rendering of his or her emotions."[34] A single physiognomic meaning constitutes the armature of any expressive face. A single analytic formula characterizes and reveals the mechanism of movement specific to each passion. But Duchenne's insistence that he did not appreciate Caravaggio (for "always finding the models of his most sublime religious scenes in gambling dens") reveals a worry: some of his photos of models with sunburned and wrinkled faces might recall the naturalism of this grand master or that of his contemporaries. Duchenne reassures his critic; he did not appreciate the new tendency in painting, either: "I am far from arriving at this modern realism that only shows us nature with her imperfections and even deformities, and that seems to prefer the ugly, the vulgar, or the trivial. On the contrary, the principles arising from my experimental research allow art to attain the ideal of facial expression."[35] Therefore one should not confuse the laws of the mechanism of physiognomy in movement with the appearance of features of subjects who serve in experimental demonstrations. Duchenne subscribes to Winckelmann's analyses on the statuary of ancient artists. Among the Greeks, study of the nude was fostered by custom, and the artist had occasion to study the play of muscles in subjects who possessed at once the strength, the skill, and the beauty of forms. The Greeks possessed mastery of that science of the living model born of observation of man in movement. But on the subject of expression Duchenne distances himself from Winckelmann. The Greeks failed in the figuration of expressive movements because they neglected the details, or because these movements were too fleeting to be observed. Duchenne had the feeling of filling a gap by showing how one might lift the rendering of expression to the level of perfection of Greek statues. If naturalism designates the imitation of beautiful nature, it also defines the application of rules of the natural language of passions. Given a naturalist interpretation of classical art, Duchenne thought he could contribute to what he calls "*ideal naturalism*"—the antithesis of Caravaggio's common naturalism. It is significant that in the aesthetic part of his book he will not forget to associate the beauty of expression with plastic beauty, in choosing a rather pretty and full-figured woman.

Resistance to Duchenne's pedagogic project and the debates he aroused are evidence of a mutation in norms. How and in what mode did the various forms of knowledge about the indicators of passions subscribe to positive notions of "norms" and "normality"? One might say that until the middle of the nineteenth century, the plastic arts are referred much more to norms than to normality. To fix the forms of expressivity, artists were invoking a visual vocabulary that designated the series of models to which they were attributing an aesthetic value. Painters and sculptors drew upon this register for a pictorial language that was pure, legible, and universal: hence the privileged relation of aesthetics to the norms constituted on the basis of an artistic experience oriented toward the classical ideal. From Le Brun to Sarlandière, passing via Lavater and Camper, the manuals and methods of drawing presented faces whose schematics expressed the essence of passions. The acquisition of mastery in the representation of expressions arose from cultural experience. By contrast, Duchenne's thinking was oriented more toward normality than toward norms: it was in relation to a type of organic structure or functioning that it enunciates its principles. A knowledge of anatomy and physiology, until then uncertain and marginal for the artist, found itself abruptly at the center of thinking on how to represent the passions. Starting with Duchenne, the art of representing the passions could no longer be the corpus of techniques of representation and the expertise that they required. Practice of the plastic arts henceforth enveloped a knowledge of the normal person. Knowledge of how indices were produced designated both an expressive function and an ordinary language. His scientific experiments led Duchenne to establish models of expression. His aesthetic thus took a normative posture, which authorized him to instruct artists in the art of painting the movements of human physiognomy. Without being the essential goal of art, imitation of nature should rely on knowledge of the normal functioning of the muscles of the face.

But the rules that govern the tracing of expressions should not be confused with those that govern human bodily proportions. The former are founded on knowledge of the grammar of the language of passions, and the latter on a judgment of taste. Duchenne did not follow his contemporaries, who claimed to establish a scientific typology of the proportions of the human body and to impose it on artists. He easily showed that it would be imprudent to want to dictate artistic norms. A canon is the formulation of a certain ideal of art incarnating an ideal of plastic beauty. Duchenne was well placed to know that a rule of proportion founded on an average tolerates deviations that are more or less marked according to different populations. The lovely instep that gives so much grace is one of the ethnological characteristics of certain races. In all ages, artists have been sensitive to such variations. But the canons that they have thereby erected arise from cultural experience. Duchenne renounced assigning a degree for the lumbar-sacral curve from an *aesthetic viewpoint*. The inclination of 63 to 64 degrees from the base and the inherent lumbar-sacral curve is the most general, but "this does not mean that this curve must be imposed on aesthetics as absolute, as a rule pertaining to the most beautiful forms. . . . Ancient art, moreover, would protest against such a law, for one finds in some statues very beautiful types of swaybacks, which were very sought after among the Greeks."[36] On the other hand, the artist cannot refer himself to the judgment of taste when it comes to representing physiognomy in movement. It belongs to the very nature of grammar to be normative, referring expression back to the normal exercise of a function. This means that the accomplishment of a function has nothing to do with the norms of fine proportion or with a fine language faithful to the rules of taste. Misunderstanding the grammar of the language of passions exposes the artist to commit, if not faults of taste, then at least faults of orthography.

This can be shown by the critical analysis of certain works of art. Lemoine did not share Latour's opinion: "It is not impossible that he was right about some of them at least. Thus he accuses the *Arrotino*, and the *Rémouleur*, and the *Laocoön* of Rome of carrying a solecism on their foreheads."[37] Duchenne followed Poussin's advice: "Read the history and the painting" [or sculpture]. Essentially, his reading retraces the avatars of the rendering of the fundamental and secondary lines. Examined separately, the frontal lines of the *Arrotino* and the form of his eyebrow are well modeled (Plate 11). But associating them is physiologically impossible. The transversal lines, which extend over the whole forehead, cannot coexist with the obliqueness of the eyebrow, because the brow and superciliary are antagonistic muscles. The result is three possible figures, because one might hesitate among three stories (read the history). One might suppose that the slave comes upon a secret conspiracy, in which case his physiognomy should express attention: the lines run across the whole forehead, but the eyebrow should be raised and rounded. Alternatively, one might suppose that the slave, upon seeing the victim whom he is ordered to torture, is seized with compassion. His face should present an expression of pain: the obliqueness and sinuosity of the eyebrows translates this sentiment, but the lines of the forehead should be limited to the median part of

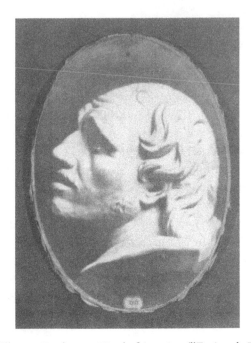

PLATE 11. Guillaume Duchenne, Head of Arrotino (l'Espion, le Rémouleur, etc.). The transverse frontal lines that extend over the width of the face can coexist, yet they are not oblique, nor are these lines formed in relation to the sinuosity of the eyebrow. Instead, they coexist due to the opposition of the frontal and the supercilliary muscles, which produce these lines, and create the oblique and sinuous movement of the eyebrow. Reproduced by permission of École nationale supérieure des beaux-arts, Paris.

the brow, and the facial planes should appear on the lateral parts. Finally, one might suppose that the slave interrupts the preparations for the torture to survey his victim: the frowning of the eyebrows should be perceived as a spasm consequent on the hurt caused by an overly bright light. With regard to the head of *Laocoön*, the median lines of the brow are in perfect accord with the oblique movement of the eyebrow. But the model of the lateral parts of the brow is impossible. The alternative is this: either the lines and reliefs that run across the brow should be continuous with those of the median part (but then they could not coexist with the sinuous movement of the eyebrow) or the obliqueness of the eyebrow that expresses pain is what is well rendered, in which case one must harmonize the brow lines with the eyebrow movements by reestablishing the natural relations between the median lines of the brow and the planes of the lateral parts. Finally, the statue *Niobe*, which Winckelmann considered the most important example of the sublime and austere style of classical art, presents no expressive sign whatever. If horror should be seizing her at the massacre of her children, it is incomprehensible that her brow and eyebrow do not present the movements of extreme suffering.

Examination of some celebrated paintings also reveals a series of faults. In *Saint François Xavier rappelant à la vie la fille d'un habitant de Cangoxima au Japon* (1641), the Japanese woman rushing to her daughter with extended arms expresses great joy mingled with pain—a discordant expression. According to the story, one might suppose that her happiness is so complete that when it first appears on her face, any lines of pain should yield to the shining of maternal joy. But Poussin does not appear to have felt this nuance: he has painted the eyebrow at its maximum painful contraction, hence the shocking contrast with the expression of extreme joy that he has given her. But there is another mistake, similar to those already noted in the *Arrotino* or the *Laocoön*: on the Japanese woman, thin and aged, the brow remains unlined despite the energetic painful contraction of the eyebrow. In this latter expression, the skin of the brow should offer some wrinkles on the middle part and some planes on the lateral parts. In certain masterpieces, it also happens that the fundamental lines are well rendered but on a single side of the face. The face of Marie de Medici in the *Naissance de Louis XIII* offers an admirable example of this discordant expression of soft maternal joy united with minor physical pain. Rubens has well represented the relief of the tip of her eyebrow and the modeling of her mouth, and her lower lid indicates maternal joy. However, something incorrect ruins the ensemble: the modeling that gives Marie de Medici's gaze a slight nuance of physical pain does not exist on the left side. The result is that in hiding the right eye, her face expresses only maternal joy. In general, the expression of cruel instincts has almost always been well rendered in the plastic arts. This is the case of the painting by Antoine Sébastien Falardeau (after Salvator Rosa), *La conjuration de Catilina*, and of Pierre Paul Prudhon's *La Justice et la Vengeance divine poursuivant le Crime*. But Paul Delaroche committed a serious mistake in his painting of *L'Assassinat du président Duranti*. He has folded transversally the skin of the frontal region, as is observed under the influence of the superciliary arch, the muscle of pain. The fundamental expressive lines of hatred and cruelty are well rendered, but the lines of pain diminish the effect of the expression of evil he wanted to give to the chief assassin.

We may set aside the discussions that opposed those loyal to Duchenne to the partisans of a totally instinctive mastery of tracing the passions. Ultimately the praise from Hubert de Verneuil, Charles Auguste Couder, Yves Guyot, and Mathias Duval made people forget Latour's reservations. In any case, the latter recognized that the rules of physiognomy in movement would be useful for students and artists in need of them. But the application of *Mécanisme de la physionomie humaine* at the Beaux-Arts has a more general meaning and bearing. It is evidence of a major transformation in the history of representing the face. Duchenne, unlike the painters who were converts to photography, was not naive enough to believe that he might be useful to artists by giving them the image of a finished painting. Rather, he wanted to respond to what artists were expecting from photography: the illumination of details that often passed unperceived or that were neglected in drawings from nature. Only a molded physiology could provide a basis for the ordering of the reliefs, hollows, and lines of the face. Duchenne's attention to tiny modifications in the face is revealing of a displacement of the site of perception of an incongruous detail. As the eighteenth century turned into the nineteenth, the representation of details was still perceived as inconvenient. One might represent old age, but one had to give it a purified image by correcting all the imperfections of nature: "It is true that, to represent age, one must imitate these degradations. But in the *details* themselves, the history painter, the artist who is occupied only with the grand, will neglect the subordinate wrinkles, the folds of the skin that in old people cross the major lines. . . . He will show what makes old age venerable, and not what heralds decrepitude: he will not afflict the spectator with the idea of destruction, and he will put before his eyes not Titon's decrepitude but the immortal agedness of Saturn."[38] With Duchenne, negation of the ideal type of old age carries an overthrow of the site of perception of what is incongruous. Anything like the image of immortal agedness, or of an idealized beauty, becomes incomprehensible. Saturn and Venus return to the museum of imaginary teratology. Realism imposes an image of old age that takes account of the physiological effects of advanced age. It little matters whether or not Duchenne's images belong to the current of realism; in any case, they are of the same age. Artists did perhaps not follow Duchenne's lesson, but it was sufficient that he had slipped his physiology of the fold into the rules of plastic art for it to belong to the history of the representation of the face.

The Frenzy of Images

"We must go back at least a century in time. Around the years 1860–1880, images were the new rage. . . . It was the completely new but shrewd, amused, and unscrupulous theft of images. Photographers made pseudo-paintings; painters used photos as sketches. A large space for play opened up."[39] In the scientific section of *Mécanisme de la physionomie humaine*, Duchenne evoked the *Laocoön* and *La chaste Suzanne*. Reciprocally, the aesthetic part made transparent the pedagogic concern that animated his preceding studies. It would not be incorrect to say that Duchenne's artistic photos give a sample of his grammar of passions. But nevertheless, should we place all his photos under the rather austere rubric of "photographic documents," as historians have done? Duchenne wanted to instruct, but he also wanted to please. In his new electro-physiological studies, the principal conditions required by aesthetics were fulfilled: beauty of form, associated with the truth of the facial expression, of attitude and of gesture. In this enterprise, Duchenne benefited from the perfecting of photographic techniques. In 1862, various kinds of progress were available: Dérosier's lens, of close focus, extremely rapid and of great depth, allowed him to avoid distortions and reduce the posing time. But the essential thing lay elsewhere: if Duchenne did not have to make a great imaginative effort to enter into the round of images, it was because it was sufficient for him to photograph not the signs of passions but their simulacra. He knew how to trace on the human face, like Nature herself, the expressive lines of the soul's emotions. We might refer to this new practice as *electrography*, since electro-physiological experimentation is like writing with electricity. But what Duchenne photographs is the image emitted by the object: the simulacrum of a simulacrum. At a time when photographers were seeking in the plastic arts the principles that would allow them to situate their practice at the level of great art, Duchenne drew these principles from his own work. The Boulogne doctor would have adopted as his own the declaration by Courbet to the Anvers congress: "These days, according to the newest expression in philosophy, one is obliged to reason even in art and to never let logic be conquered by sentiment. Reason should be in everything man's dominant characteristic."[40]

It is useful to recall how Duchenne modified the relations between the operator and the model, as well as between the spectator and photography. What had been in the scientific part of his book the study of an expressive function becomes here the dramatization of passions, in which he effects a first break with photographic practice. At the time, practitioners expected from a model either voluntary mimicry or an expressive movement that spontaneously crossed the face. The initiative still lay with the photographic subject. By contrast, Duchenne expects nothing from his model except that he or she remain passive. Hence the choice of a large young woman, rather well proportioned and whose facial traits were regular. But being almost blind, she could not see the attitudes that he was indicating to her, and so he was forced to pose her and drape her as if he were dealing with a mannequin. An artisan of expressive forms, Duchenne first engages his model in performing a few facial movements that contribute to shaping them. He fixes the angle of her head, the direction of her gaze, the closing or widening of the eyelids, the greater or lesser opening of her mouth. If Duchenne begins by presenting a portrait of his model, it is in order for people to see natural features as they appear in repose. Occasionally he would integrate them into the composition of a more complex expression. Thus in this sequence of portraits we see the natural lowering of the corners of the lips, which accentuates the feeling of sadness or the expression of aggressive passions. Then, by the electrical stimulation of muscles that contract only involuntarily, the experimenter obtains the sought-after expression.

The second break concerns the relation of the spectator to the image. Photographic plates demand no effort of comprehension. They are meant to arouse in the person looking at them emotions similar to those that he might feel in front of a painting. Duchenne expected a lot from the onlooker, and this onlooker was the required counterpart for the new rules of reading he was proposing: nothing less than a technique of looking. In effect, electrography implies a frontal presentation of the subject and places the observer in front of aberrant expressions. So it is necessary to introduce masking in order to cover one part of the face so that the completely drawn facial expression may be seen. This arrangement, articulating electrography and photography, determined a series of modifications in the field of *picturesque* photography. Finally, here is the story of a little-explored moment in the history of artistic photography.

The first path taken by artistic photography was in the direct lineage of the expressive portrait. Photographers were applying the precepts recommended by painters. Essentially, this meant sketching an expression taken on the fly, drawing or painting a living model, and, by skillful composition, aiming for beauty without losing any of the truth of the subject. At first, photographers tried to capture expression from nature. For example, for the expression of weeping or shame, the disposition of the mouth is well represented in two photos by Oscar Gustave Rejlander that Darwin reproduced in his book. In these figures, we see one young boy who has just stopped crying, while the other boy is on the point of bursting into tears (Plate 12)—it is at this precise moment that Rejlander has chosen to photograph them. Practitioners might also ask their model to mime an emotion. It was for Darwin, too, that Rejlander photographed a woman who could simulate defiance and men who mime resignation, disgust, and astonishment: the gesture reinforces the expression. Finally, photographers made portraits in the manner of painters, endeavoring to express the temperament, grace, and physiognomy that constituted the character of their subject. But a fleeting expression is only one of the elements that contribute to the success of an idealized portrait; others include posture, clothing, and accessories. To grasp what is essential without lapsing into exaggeration, there were only two solutions: either to represent a face conforming to the traditional schema or to give an image based on a scrupulous and selective realism. In the first case, practitioners were returning to academicism. Rejlander recognized his debt to Le Brun, and it was after the latter that Henry Peach Robinson made a series devoted to the emotions, *The Passions* (1857). The idealized portrait of Jean Journet by Nadar, *The Apostle* (1855–1856), also recalls the expression of ecstasy by Le Brun: head turned upward, eyeballs drifting back, and mouth open. In the latter case, we rediscover one of the precepts taken from painting: it is impossible to use everything, and so you must choose the expression that completes and complements the resemblance of the portrait. "It is always the same rule that should direct the artist; it is absolutely necessary that through all the passing nuances of the face, all these airs of the head, these smiles, these fleeting movements of lips and eyes, that he untangle the right and eloquent expression that contains, more than all the others, the individual

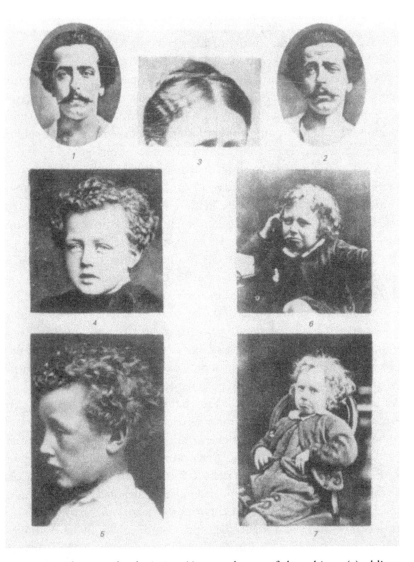

PLATE 12. Photographs depicting (1) natural state of the subject; (2) obliqueness of the eyebrows in the expression of pain; (3) transverse wrinkles of the face in
pain; (4) natural expression; (5) obliqueness of the eyebrows in expression of pain
before melting into tears; (6) ceasing to cry; (7) contraction of triangular and lowering of the corners of the mouth. From Charles Darwin's *L'Expression des emotions*
(Paris, 1890), table 2. Reproduced by permission of Bibliothèque de l'académie
nationale de médecine/Université René Descartes, Paris.

character, what summarizes, in a way, all the various manners of being of his face."[41]

In his electro-physiological studies, Duchenne had extracted the laws of the expressive mechanism. The epistemological precedence of his images involves a first transformation of the relations between photographs and paintings. What Duchenne demands of the teaching of the plastic arts are not precepts but subjects for competition. To be precise, the prize for the expressive head competition always counted among the most prestigious and best endowed at the School of Beaux Arts. In 1861, one of the competition's subjects was "Contentment of a Mother after the Healing of Her Child," and so he did a photo that responded to that year's subject. "Certainly nothing would be easier to portray than the joy of a mother who senses that her child is coming back to life. We confirm that this expression is fairly well rendered in figure 79, if we cover the left eye."[42] But Duchenne is interested above all in combined and discordant expressions that put into play contrary muscular contractions. They are more difficult to render than the preceding expressions, as already shown by the critical analysis of Poussin's painting: Poussin did not know how to represent the expression of joy mingled with pain. To really capture this sentiment, the median electrical contraction of the superciliary muscle (muscle of pain) has to be associated with the expression of joy: "In the scene comprising the principal subject of figure 79, the expression is complex. Actually, the happiness of this mother whose last child has just escaped death could not have made her forget so soon the one who has just died. Her maternal heart is therefore seized at the same time by two contrary emotions: joy and pain. It is this that is expressed in the left half of Plate 79, when the right eye is masked." Figure 80 also illustrates a subject for the competition, "Compassion" (1853). But Duchenne distinguishes two nuances in this state: on the one hand, benevolence is expressed by a simple tender smile, which translates a disposition favorable to others; the accent is on a form of condescension without any pejorative connotation. But on the other hand, compassion is what results from the association between light electric contraction of the small zygomatic (muscle of moderate crying) and the natural smile. This expression is subtler than that of benevolence: it translates a sentiment that tends to share the suffering of others. Moral virtue linked to the idea of suffering is incarnated in the compassionate smile: "The young woman photographed in this figure is visiting a poor family; we recognize, from her tender smile (cover the left side of the face), or from her kind smile (cover the right side of the face) that she is touched by the misery and the suffering of this unhappy family, and that this sentiment has inspired an act of charity."[43]

It is useless to extract the implicit philosophy that underlies this vision of the social world, for Duchenne's interest in maternal feelings and in charitable women does not exhaust the meaning of his photos. His goal is not to elaborate expert compositions, but to show specimens of expression; the same expressions would suit other situations that might inspire the sight of these figures. Hence there is displacement of the problem of representation of complex expressions in relation to the practices of expressive portraiture. As long as one was reduced to applying the recipes of the plastic arts, the register of representable expressions was limited. The expressions that could be captured on the fly, on condition that they are readable, are crying, chagrin, and laughing. The other expressions that might be seized thanks to a model gifted at imitation, under the same condition, are those found in plates 4, 5, 6, and 7 of Darwin's book: defiance, disdain, disgust, indignation, and astonishment. Shooting them presented difficulties that the photographers resolved by different tactics: gently, through Lewis Carroll's infinite patience, or forcefully, through Julia Margaret Cameron's impatience, as when she shut her model in a cupboard to obtain the expression of despair. To seize these fertile moments, one needed to observe each instant and have careful tact. Finally, if one wanted to move beyond the pale copy of the model and capture "the soul itself" (André-Adolphe-Eugène Disdéri), or "the intimate resemblance" (Nadar), one had to combine photographic practice with the exercise of a pure sensibility. We thus re-encounter the norms of an art that could not be learned. Photographic practice is not summarized in counsels of perseverance, skill, and taste: "This art of portrait making can only be acquired by a constant observation of nature and by long and patient work. Still, these studies will not be fertile except for an artist naturally endowed with the sentiment of beauty."[44] From the moment Duchenne masters the art of summoning the signs of the passions on the face, he is able to re-create the series of prototypes of passions and their nuances without needing to resort to the tactics of photographers. The problem of the fertile moment disappeared, since the acme of the expression coincides with the contraction of the expressive muscles. And strategies founded on aesthetic experience and sentiment were no longer relevant. It sufficed to apply the rules of the mechanics of the face in movement. For the subjective criteria of pure artistic sensibility, Duchenne substitutes the photographic recording of expressions that obey the rules of harmonious representation.

The second path pursued by artistic photography was in the direction of the reproduction of works of art. Very early on, Joseph Nicéphore Niepce had sought the means by which light might facilitate the transfer of a draw-ing onto another medium, which would permit reproduction in series. Starting in 1839, Hyppolite Bayard issued a series of photos of engravings and paintings. *The Pencil of Nature* (1844), by Fox Talbot, is the first book illustrated by photos reproducing works of art. Shortly after, Blanquart-Evrard edited his *Album photographique de l'artiste et de l'amateur* (1851), including reproductions of drawings, sculptures, engravings, and paintings. Another example is *L'Oeuvre de Rembrandt, reproduit par la photographie* (1853–1858) from the Bisson brothers. Very soon, practitioners shifted from reproducing a painting to the reconstruction in the studio of a scene that recalled this picture. They found their motifs in the history of painting and subjected their own images to its canons. Photographic practice referred back to the composition of the *tableau vivant*. In the context of re-creating famous works of art, Rejlander founded his popularity on a photo that evoked the arrangement of *L'École d'Athènes* by Raphael. In *Les deux modes de vie* (1857), the composition is governed by the constraints of classical painting. Photography, as previously noted by Francis Wey in reference to calotypes, excelled in the imitation of the most individual styles: in a land-scape series, he saw examples of Maurice Joyant, Giovanni Battista Piranesi, Jean-Baptiste Camille Corot, J. I. van Ruysdaël, Prosper Marilhat—all for-tuitously bursting from the sole fancy of Nature. It was by a determined and scrupulous imitation of the motifs of academic painting that Cameron pro-duced images that recalled the style of a painter. Openly taking inspiration from the masters of the Italian Renaissance, she composed images in the styles of Francesco Francia, Pietro Perugino, and Raphael: *Saint Agnès* and *Beata Beatrix* (1864). She also made a portrait of a child (*Waiting*, 1872) in which the pose and expression are taken from a *putto* in Raphael's *Madone Sixtine* (1541). Finally, photographers applied themselves to reconstituting in the studio scenes similar to those demanded of a painter as exercises. Scenes taken from nature in the studio, as well as poses by artists' mod-els, demonstrate an imitation of academic "clichés." The idealized forms of academic painting were an inexhaustible source of inspiration. The use of nudes would contribute to the integration of photography into the Beaux Arts. Ingres was widely put to use: the nudes of Jacques Antoine Moulin, for example, are the direct descendants of the Oriental odalisques and Venetian Venuses.

On the basis of his experimental studies of the passions, Duchenne had elaborated a dictionary of expressions. The iconographic precedence of his images carries a second transformation of the relations between photography and the plastic arts. Duchenne turns toward certain works of art not to find his motifs in them but to see if the representation of expressions was accurately captured in the tracing of expressive lines. He did not want to reconstruct a *tableau vivant* by taking inspiration from Gian Lorenzo Bernini, Guido Reni, or Ingres, nor even copy their styles. If he makes reference, for example, to a group of statues by Bernini that he saw in Rome, it is in order to stress the misconceptions. Was he depicting the ecstasy of heavenly love (the expression that ought to be read on the face of Saint Teresa)? Then Bernini did not know how to show it: "Bernini's group represent[s] the ecstasy of St. Teresa . . . a beautiful angel armed with a spear appears to her and everything in [her] expression breathes the *most voluptuous* beatitude."[45] But her actual expression does not conform to the sentiment that ought to be perceived on her face. To render the traits that correspond with the concept of mystical ecstasy, it would have been necessary to show what appears only on the left side of figure 77: the soft rapture of divine love (smile, half-opened mouth, and a slight obscuring of the eye, obliquely turned upward and slightly laterally). But Saint Teresa's expression becomes irreproachable as soon as Duchenne indexes it on the right side of figure 77, thereby showing the ecstasy of terrestrial love (moderate stimulation of the nose transversal).

There is more than meets the eye precisely because there is less: the photos that Duchenne did not take—not because the rationale of high aestheticism requires a separation between high art and the vulgar images of cynical ecstasy, but because the latter would have revealed the staging of a fantasy. Duchenne's shock at his experimental audacity shows a disarming naiveté. However, we can only admire the skill with which he was able to integrate the technical constraints of localized electrical stimulation into his *vision*. How this fantasy is evoked is splendid: the appearance of the operator in photography. The archangel Gabriel has given way to the experimenter. Ready to work, he holds in his right hand the rod of the stimulator, an arrow or small brush that is going to pierce the heart of Saint Teresa. Following the wand, one perceives that the electrical fluid must have been unflinchingly launched on the muscle of lasciviousness and that it is one body with it, or rather, with them: "Therefore, I have performed a meta-morphosis by changing the purest, most angelic smile and the most saintly ecstasy of figure 76 into the most provocative and licentious by means of combining a strong contraction of this muscle with the other features; in so doing I transformed virgins into bacchantes."[46]

Of course, the concept of voluptuous lasciviousness springs from the idea of woman as a power of debauchery. But this theme does not exhaust the meaning of Duchenne's aesthetic photographs. He reminds us that several figures in the scientific part of the album had already been devoted to the study of the muscle of lasciviousness, which is activated under the influence of amorous excitation, the pleasures of love. These are the faces of men whose expression is similar to that observed in the figures of satyrs and fauns. Duchenne believes that this simple, sober, and elo-quent facial movement expresses sexual excitement. Hence there is a dis-placement of the problem of the harmonious representation of expressions in relation to the practices of servile imitation. As long as photographers sought their models in painting, they were confronted with the problem posed by the representation of a clear and readable expressive language. To palliate the extreme paucity of facial expressions representable by this medium, practitioners had only two solutions. The first was to accentuate a purely formal resemblance to the "looks" [*les airs de tête*] by the inven-tion of technical artifices: retouching the negative or the print in order to achieve an idealized reality. The second solution was a surfeit of decorum, meant to contribute to facilitating the reading of the image. This prac-tice was put into effect both in the production of an image in the style of a painter and in studies of nudes. The chubby but inexpressive child who serves as model for Cameron recalls an angel with cotton wings, and so one is more reminded of a costumed child in a pageant than of Raphael's *putto*. In studies of nudes, the woman is always seductive. But the deco-rum, far from appearing as an element in the composition, instead evokes a maladroit transposition of the language of academic painting. From the moment when Duchenne is in a position to grasp the characteristic traits of the passions, he excludes the techniques that were supposed to enable overcoming the medium's limits of representation. Manual intervention to perform retouches would be incompatible with its goal: the representation of natural expressions. As for the staging, it could now opt for sobriety. In the representation of terrestrial love, the young woman's hair falls to her bare shoulders and her opened hands are brought to her chest. The half-dressed state and passive attitude complement the expression, already readable, of lasciviousness.

Illustrative painting had opened up the third avenue taken by photography. Photographers thought they could rival sketch artists or engravers by responding to publishers' demands. From the reconstitution of a real painting, practitioners moved to the composition of a *tableau vivant* illustrating a poem, a legend, or a theater play. Cameron illustrated the *Idylls of the King*, Tennyson's poetry collection inspired by Arthurian legends. Robinson issued the very popular series *Little Red Riding Hood* (1858). The painter-decorator William Lake Price illustrated Shakespeare, Defoe, and Cervantes. Roger Fenton, a painter by training, illustrated the *Seasons* by James Thomson and Ronald Leslie Melville did *Ivanhoe* and *Talisman* by Walter Scott. Photographers also arranged their scenes so as to reproduce images that recalled the style of a school of painting, in particular the Pre-Raphaelite School founded by the painters Dante Gabriel Rossetti, William Holman Hunt, and John Everett Millais. The links are evident in the photos of Cameron and Robinson, with the former presenting sacred subjects, biblical figures, and medieval legends, while the latter drew his subjects from Tennyson's *Idylls of the King*, in a version of *Elaine with Lancelot's Shield* (1859–1860) and the famous *Lady of Shallot*, adapted from Millais's painting representing Shakespeare's *Ophelia*. Of this collusion between photography and painting, John Ruskin bears indirect testimony in his defense of a school unjustly suspected of plagiarism: "When . . . it began to be forced upon men's unwilling belief that the style of the Pre-Raphaelites was true and was according to nature, the last forgery invented respecting them is, that they copy photographs."[47] But it was Baudelaire who gave the death stroke to illustrative photography. It was a crude lie: "By bringing together and posing a pack of rascals, male and female, dressed up like carnival-time butchers and washerwomen, and in persuading these 'heroes' to 'hold' their improvised grimaces for as long as the photographic process required, people really believed they could represent the tragic and the charming scenes of ancient history."[48] In the study of nudes in the academic tradition, photography was sliding toward pornography. This social phenomenon also aroused Baudelaire's indignation: "A little later a thousand hungry eyes were bending over the peepholes of the stereoscope, as though they were the sky-lights of the infinite. The love of pornography, which is no less deep-rooted in the natural heart of man than self-love, was not to let slip so fine an opportunity for self-satisfaction. And do not imagine that it was only children on their way back from school who took pleasure in these follies; the world was infatuated with them."[49]

In his study of facial expression, Duchenne had renewed the art of photographing facial expressions. The guiding light of his images carries a final transformation of the relations between photography and the plastic arts. He turns toward the graphic arts, not to borrow their subjects to be illustrated in the style of any school or even in his own way, but in order to find the principle of narrative sequence. It is not by chance that Duchenne cites at length the "Essai de physiognomonie" (1845) by Rudolphe Töpffer, inventor of print-illustrated literature. A succession of images that tell a story presupposes a series of formal resemblances between photos and the drawings of a caricaturist: richness of facial details, precision in the indicators of emotions, the eloquence of a story-in-images that can be widely understood. In Töpffer's lineage, Duchenne invents the photo-novel in the form of a series of three images that represent the nuances of a passion and illustrate a story. Figure 81 shows "Lady Macbeth: *Had he not resembled my father as he slept, I had done 't.* Moderate expression of cruelty. Weak electrical contraction of the nasal pyramid." Figure 82 (Plate 13): "Lady Macbeth: *Come you spirits . . . / And fill me from the crown to the toe full / Of direst cruelty.* Strong expression of cruelty. Medium electrical contraction of the nasal pyramid." Figure 83: "Lady Macbeth at the moment of assassinating King Duncan. Strong expression of cruelty. Maximum electrical contraction of the nasal pyramid." No less than three photos: this muscle is situated on the median line of the face; the play of masking [*caches*] is impractical here.

Another photo (figure 78) presents a series of shots that tell a small story. Duchenne narrates various moments in a scene of seduction. This time, he fixes his fantasy by capturing what he sees through the keyhole. The operator surprises his coquette at her toilette (Plate 14). An actor in the role of an audacious seducer (the experimenter in the photo), his two hands

PLATE 13. "Come, you spirits / That tend on mortal thoughts! unsex me here, / And fill me from the crown to the toe full / Of direst cruelty" (Shakespeare, *Macbeth*). "Lady Macbeth" in an expression of intense cruelty from a moderate electric contraction of the pyramidal of the nose. From Guillaume Duchenne's personal album, figure 82. Reproduced by permission of École nationale supérieure des beaux-arts, Paris.

are posed on her shoulders. The taking is assured in a three-shot sequence: offense, disdain, mockery. First, the offense: "A gentleman surprises a young lady while she is dressing. On seeing him her stance and her look become disapproving (cover the bottom half of her face)." Then disdain: "The mannered pose of her hand, which supports a rather overtly revealed bosom. The young man was becoming more audacious, but the words 'Get out!' pronounced in a scornful way by the girl, stop him in his enterprise (see only the left side of the lower half of the face)." Finally, mockery: "The mocking laughter that accompanies the amorous rejection . . . we believe to mean 'Conceited ass!' Perhaps she says also, much lower: 'The fool, if he had dared . . .' (see the right side of the lower half of the face, electrical stimulation of the grand zygomatic, eyelids slightly closed, and gaze directed slightly laterally)."[50]

Using these conventional themes, whether illustrating the dialogue of Shakespeare's play or a scene of coquetry, Duchenne reveals his variety and finesse with expressions that themselves tell a story. There is a displacement, then, of the problem of how to sequence a story from how to choose a scene

PLATE 14. Scene of flirtation, with different expressions on the right and on the left sides of the face. From Guillaume Duchenne, *Mécanisme de la physionomie humaine* (Paris: Renouard, 1862). Reproduced by permission of Bibliothèque de l'académie nationale de médecine, Paris.

taken from a story. As long as photographic practice had no other purpose than to illustrate a narrative moment—whether to pastiche a school of painting or to produce a suggestive image—it accommodated a series of immobile expressions. All these tableaux are nothing less than living. In passing, note that the vogue for images representing subjects who are either sleeping or on their deathbeds—Courbet's painting *Sleep* (1866), Henry Peach Robinson's composite photograph *Fading Away* (1868), and Amy Scheffer's painting *Felicité-Robert de Lamennnais écrivan* (1854)—bears indirect testimony to the powerlessness of photographers to represent the passions—as in the example of pornographic photographs. In the many anonymous nudes (daguerreotypes of the 1850s), but also in the nudes of Moulin and Bruno Braquehais, we always find the same cajoling woman, provocative and cloying. We also find the same gesture: hand brought to the height of the face, thumb under the chin, and index finger against the cheek. The same question, or rather, the same wink at voyeurs. These scenes of seduction illustrate more than they legitimate the idea that working-class girls are interested in only one thing. But there was not enough there to *make a story*. From the moment Duchenne applies his knowledge of expression in his aesthetic photos, the image tells a story, since it incorporates several ideas in sequence. Of course, to make the image speak and in order for it to seem animated, one would have to slip in the masking (Plate 15). We are still far from the sequential photography of Eadweard Muybridge or Étienne-Jules Marey, but it is a first push in that direction, and its consequences should be underlined. All the significance of the story is read on the face, which no longer appears as the poor spokesman of an incitement to debauchery: the spectator is no longer even interpolated. The message is subtler than that: everything indicates that this coquette was feigning indignation but that at bottom she was really flattered by her audacious suitor's enterprise. In allowing the woman's thoughts to be seen and to be read, the good doctor did not suspect that he was offering the most eloquent image of phallocentrism. This is no longer the crass speech of pornographic photos where the girls always say yes. The coquette always says no, but she thinks yes.

Duchenne's contemporaries debated considerations about the utility of the studies on expression for the teaching of the plastic arts. Yet his aesthetic photographs were not the subjects of any commentary whatever. If this series went unperceived, it was because it was confused with the scientific photo series. How did this confusion come about? Without

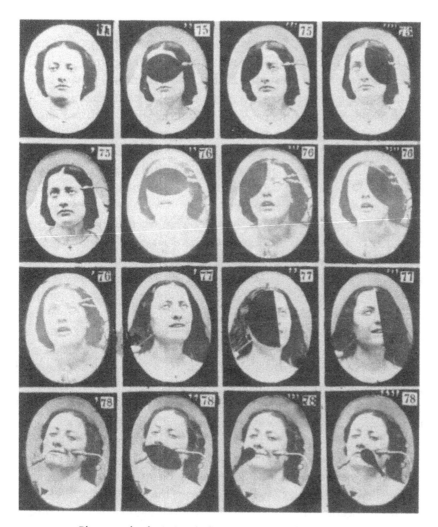

PLATE 15. Photographs depicting (74) woman at rest; (75) nun pronouncing her wishes, with resigned pain on the left and only sadness on the right; (76) taking the veil, with profound pain on the left and divine happiness on the right; (77) terrestrial love on the right and celestial love on the left; (78) scene of flirtation. From Guillaume Duchenne's personal album. Reproduced by permission of École nationale supérieure des beaux-arts, Paris.

entering into debates about the status of photography, one might say that people agreed in that era on a distinction between two types of images. On the one hand, they asked scientific photography to show the hidden aspects of reality. Photography was an auxiliary to microscopy or astronomy in showing what escaped the naked eye. There was also its documentary role: for drawing up military plans, the study of plants, animals, and the human races. On the other hand, people wanted artistic photography for the dream aspects it carried. It was assigned the status of a work of art as soon as people could identify in its images the effects of an idealized reality—in particular, the immemorial function of the portrait. This is not surprising, since pictorial practice was among the conditions of the emergence of artistic photography. Very schematically, we may say that these two functions of the medium were perceived as mutually exclusive. But the interweaving of both genres marked Duchenne's aesthetic photos. The critics' silence relates to the fact that his photos are unclassifiable, since they are unique in their genre. Historians of photography have hardly been any more forthcoming. Some have mentioned an aesthetic of ugliness or even cruelty. Others have compared his images to the caricatures of Honoré Daumier or the paintings of Théodore Géricault. Still others have seen in them various kinds of allusions, sometimes to academic painting, sometimes to commercial photography, and sometimes even to aspects of religious art. Duchenne's photos are taken as poor testimony of the era. All these comparisons are perhaps not false, but they are hugely inadequate. They do not take account of the singular place of Duchenne in the photographic production of his age. The Boulogne doctor captured the transformation of the face: his model presents so many new appearances that it is hard to believe that it is always the same model. One might say that by electrography, the model's face is transfigured, which means, in particular, transformed by giving to the face an unusual brilliance—flashes of joy, happiness, suffering, sadness, cruelty. If it is necessary to put a label on these images, one would have to say that at a time when theatrical photography was triumphing, Duchenne was inventing the theater of affects: *transfigurative* photography.

Conclusion

"*Descartes' Error*" is a title that might well produce Heraclitus's tears or Democritus's laughter, according to whether one is prone to good or bad will toward those who commit anachronisms. But Antonio Damasio at least tells a sympathetic story:

My mentor Norman Geschwind, the Harvard neurologist whose work bridged the classical and modern eras of brain and mind research in humans, was fond of pointing out that the reason we have difficulty smiling naturally for photographers (the "say cheese" situation) is that they ask us to control our facial muscles willfully, using the motor cortex and its pyramidal tract. (The pyramidal tract is the massive set of axons that arises in the primary motor cortex, area 4 of Brodmann, and descends to innervate the nuclei in the brain stem and spinal cord that control voluntary motion through peripheral nerves.) We thus produce, as Geschwind liked to call it, a "pyramidal smile." We cannot mimic easily what the anterior cingulated can achieve effortlessly; we have no easy neural route to exert volitional control over the anterior cingulate. In order to smile "naturally," you have only a few options: learn to act, or get somebody to tickle you or tell you a good joke. The career of actors and politicians hinges on this simple, annoying disposition of neurophysiology. . . . The difference between facial expressions of genuine and make-believe emotions was first noted by Charles Darwin in *The Expression of the Emotions in Man and Animals*, published in 1872. Darwin was aware of observations made a decade earlier by Guillaume-Benjamin Duchenne about the musculature involved in smiling and the type of control needed to move that musculature. Duchenne determined that a smile of real joy required the combined involuntary contraction of two muscles, the *zygomatic major* and the *orbicularis oculi.* He discovered further that the latter muscle could be moved only involuntarily; there was no way of activating it willfully. The involuntary activators of the *orbicularis,* as Duchenne put it, were "the sweet emotions of the soul." As for the *zygomatic major,* it can be activated both involuntarily and by our will and is thus the proper avenue for smiles of politeness.[1]

Thus it happens that biologists recognize the pertinence of the observations made by the doctor from Boulogne. A small homage would have been to render what belongs to him: his smile, "the smile of Duchenne." Historians of ideas, more concerned with radical breaks, mention Darwin as the starting point: "The natural history of the expression in the 19th century is marked by a major break; when in 1874 C. Darwin published *The Expression of the Emotions in Man and Animals,* the study of facial expression was transformed. . . . Human expression is now grasped on the basis of experimental observation, but also derives from a continuum that puts humans in relation with their animal origins: at the very source of the expression of emotions, there no longer lies language but the organism."[2] But things are not that simple. Underneath the evidence of a major break are found the historical and concrete conditions of an epistemological reorganization, which had the effect of redefining the significations of the physiognomy in movement. Duchenne, who believed in the existence of an original language of the emotions, could not pose the problem of the genesis of facial expressions. By contrast, Darwin did perceive emotional language as a behavior linked to a determinism and to a history. But Darwin could only do his study on the basis of a redistribution of knowledge, which took effect through Duchenne's work. Before facial expressions could become a special branch of comparative biology, they first had to be recognized in their biological dimension, as the expression of a function. Until then, the language of the passions had been studied only in its semantic structure. Classical analysis had given facial expressions the same status as speech. Having misrecognized the fact that the language of affects depends on a mechanism, the psychology of passions remained at the level of a metaphysics of the senses. The expressive act itself was never reconstituted in its necessity. Within physiognomic movement, Duchenne recognized a natural language that exists with all the rigor of its grammar. To express one's passions, it is necessary to possess in one's anatomy and in one's physiology a structure of expression. Producing a facial expression presupposes an expressive mechanism that allows it to give body to what it expresses. The language of passions then ceases to be transparent to the representations of any culture and thickens as it receives its own weight. But in order to fit expressive acts with the organism, Duchenne was obliged to admit they had a purpose. In pushing back the innateness of involuntary expressive actions onto a pre-determinism, he was reaching an impasse. Discovery of the grammar of this natural language was inseparable from a theology, but there is nothing in this relation that annuls the positivity and force of his analysis.

Historians of science have made astonishing discoveries. Quite recently, Robert Sobieszek indicates where *Mécanisme de la physionomie humaine* ought to be situated: "Scientifically, it is locatable somewhere between Lavater's work on physiognomony [*sic*] in 1775–78 and Darwin's on expressions in 1872." However, Lavater's physiognomy was a taxonomy of characters that excluded a pathognomy. And the study of the face in movement could appear only when it was based on the recognition of the ensemble of anatomical and physiological structures of muscles and skin. Knowledge about expressive acts that escape voluntary control was what made possible Darwin's analyses on the expression of emotions. No doubt fascinated by the method of localized electrical stimulation, Sobieszek was able to write: "In medical terms, Duchenne's *Mécanisme* is poised between Giovanni Aldini's attempts in 1802 to reanimate cadavers with galvanic shock and the introduction of electroshock therapy for psychiatric patients by Ugo Cerletti in 1938."[3] But the application of the same procedure does not always carry the same meaning. There is nothing in common between the therapeutic fantasies of Aldini and Cerletti and the exploration of the expressive function. *Mécanisme de la physionomie humaine* is a book of physiology, not a manual on electrotherapy. Strictly speaking, if it relates to medicine at all, it is by integrating facial pathologies so as to elucidate the tonic function of muscles and by a clinical approach in the invention of a symptomatology of the passions.

The history of the discovery of the expressive field held many surprises. Descartes had associated the signs relating to symptoms with cardiopathy, and he placed facial expressions as depending on the will. It was necessary to await the neuro-physiological studies of Charles Bell for facial expressions to become symptoms and to find their anchor point in the respiratory function. Later on, Duchenne shifted the substratum of expression from the respiratory organs to the facial muscles. By the substitution of a myopathy for a pneumopathy, he produced an ultimate shift. From this, we may see why Duchenne was interested only in expressive acts that arise from the play of muscular actions. We also understand why he left in the shadows any other functions of facial muscles. Duchenne could not help being indifferent to the fact that there are no organs specifically designed for expression. The question of the diversion of the specifically physiological uses for expressive ends was not his concern. Muscles are quite naturally the instruments of expression. From this angle, Duchenne's perspective is rather that of a psychologist.

Curiously, the representatives of the discipline to which he brought a major contribution were not any more clairvoyant. Georges Dumas has said, "It would have been better for a great clinician like Duchenne not to do philosophy."[4] That a psychologist remains blind to the philosophical questions that border on the scientific foundations of his discipline should not surprise anyone. Duchenne belongs to that type of philosopher whose research passes for esoteric because they give the impression of complicating what seems to go without saying, whereas they are inventing a new problematic. Dumas does not see that what is recognized without an analytic effort, for this very reason, constitutes an original problem. As Ludwig Wittgenstein said: "The aspects of things that are most important for us are hidden because of their simplicity and familiarity. (One is unable to notice something—because it is always before one's eyes.)"[5] Before a face, one feels an overall impression. One perceives a tonality and not the muscular contractions, changes in dimensions, or shifting lines. This perception does not raise any issues since it is immediately grasped by the senses. The power of this mode of expression, its transitory character, and its habitual usage contradict the contemplative attitude.

In the history of emotions, one has rarely measured the weight of everything it would be necessary to leave behind in order to discover the expressive field. Interests of an ethical, cultural, and aesthetic order that gravitate around the passions would have to be set aside. Here we will retain only those that are linked to the phenomenon of expression. Merely grasping the meanings of various expressive contents passed for a long time as a kind of knowledge. This is why the first commentators did not hesitate to situate Duchenne's work in the lineage of physiognomy: "Applied to a particular action of the muscles of the face, he developed through electrophysiology all the mechanical phenomena of physiognomy upon which, as we know, Lavater had made his ingenious studies."[6] If it sufficed to perceive the expressions in order to know them, how can we explain that people never knew how to *describe* them? Until then, the given task was to interpret them, to determine their expressive value, and to know whether they spoke true or false. The movements of the physiognomy were so many fugitive traces that had to be made to speak. Duchenne shifted the point of analytical attack: the expressive act and the act of recognition become the objects of knowledge. For the first time, the fundamental questions are posed: What does expressing mean? What does recognizing it mean? From now on, it is a matter of working on expressions from the inside and from the outside, simultaneously. It is a matter of analyzing them, finding their elements, describing their relations and their rules of composition.

Again quite recently, Anne-Marie Hans-Drouin has reactivated Gratiolet's objection: "Duchenne ended up with the idea of a universal vocabulary of expression, but by a means in which, precisely, no emotion is 'felt.' . . . May one speak of the expression of the soul when only the body participates in muscular modifications?"[7] By repeating the old objection from the spiritualists that Duchenne put the soul between parentheses, one has ended up losing sight of the formal beauty of his experiments. It is on the basis of simulation that he was able to proceed to the first transformation in the knowledge of expression. It was necessary to really *see* the photography that he made of a specimen of his electro-physiological experiments. The experimenter applies to his model the electrodes of his machine, but without sending any current (Plate 16): "I must say that in this case the laughter is natural! I merely wanted to show a simulation of one of my electro-physiological experiments."[8] This image summarizes the epistemology of the doctor of Boulogne: a natural facial expression in the place of a provoked one, and vice versa. This means that the modes of releasing expressions are different but the expressions produced are identical. On the one hand, the presentation of a provoked expression is the duplicate of a natural expression. In the relation between the physiological mechanism and the *sur-face* language, nothing is lacking. In the action of the electrodes, in the same way it proceeds from an affect, it is the same structure of indication that appears. On the other, Duchenne is aware that the artificial expression, although it might be confused with a natural one, differs fundamentally from it. By artificial expression, the experimental subject expresses a way of being without being himself affected. By the natural expression, the

PLATE 16. Specimen of an electro-physiological experiment. From Guillaume Duchenne's personal album, figure 1. Reproduced by permission of École nationale supérieure des beaux-arts, Paris.

person expresses both a way of being and a way of being himself affected—a behavior. But Duchenne's interest is limited to the surface. One might say that there is no passion without expression, and that the former without the latter is nonsense. But one cannot say, reciprocally, that there is no expression without passion and that a provoked expression is nonsense. In any case, whether they are accompanied by an affect or not, we perceive only the images of passions. Duchenne stresses the opposition between the expression and the absence of affect in order to turn it to the benefit of his description. The discourse of the surface is opposed to the emptiness of the soul, but it cuts across the language of passions. The concept of "passion" surfaces and becomes the soul of the image. This discovery provided the key to deciphering the *mechanism* of the face in movement: "I will designate this kind of research to which the soul remains entirely foreign, under the title 'Mechanics of Physiognomy' to indicate that I make use only of a physical force to produce this peripheral muscular play."[9] The analysis of muscular actions, simple or combined, refers to the basis of the grammar of physiognomy in movement. And this natural language of the passions has less the sense of "speaking" than the original one of *dis-courir* [*discourse*]— spreading, going or running here and there, under the skin and on the skin. Lewis Carroll, also a photographer, said, "The surface plane is the character of a discourse."

But what then, to us, is this discourse of the surface? For the seen reality of expression, referring to the whole movement of the face, Duchenne substituted the re-seen reality of an optical illusion. He compared the effects of the passions to the effects of an illusion performed on the visual organ by juxtaposing certain colors. The mirage that we are made to feel by certain circumscribed movements of the face is of the same order as that kind of optical illusion exercised by the simultaneous contrast of colors. Hence he aligns the concept of perceiving such indices with the concept of perceiving colors. If agreement can eventually be reached in the language of those who perceive a facial expression, this is because they first agree in an error of judgment that is committed quite naturally by those with normal sight. "When we see a limited movement and recognize the perfect image of an emotion, it seems to us that the face has changed in an overall way. When we experience such illusions, it is by virtue of our organization, by virtue of a faculty we have possessed since birth."[10] Just as the recognition of the necessity of illusions forces us to cease equating being with appearance, so we must renounce equating the expressive act with a *ready-made* expression. This means that the spontaneous language of passions is the opposite of the voluntary expressions that are given as silent speech. We must not confuse natural language with mimicking language, the grammar of physiognomy in movement with the linguistic grammar. People do not associate the image of an emotion with perception of a face in movement, but they recognize right away the image of an emotion. To recognize is to link the present image with that which has already been perceived, or else to find again in the present image what was already known. To be significant, facial expressions must not only be founded on real behavior but also be perceived in that very moment. They are above all behavioral, and on occasion, a form of communication for someone who recognizes their structure of indication. Mimicked expressions do not communicate anything. The spontaneous laugh, index of joy, is not a play of language and can never become so, whatever efforts people might deploy in order to imitate it. The false laugh is an authentic sign that, unlike the index, does not carry the joyful affect. With the feigned laugh, the mimic does express something, but it is no longer a form of expression since it topples over into the language of mimicry. It is another kind of behavior: a form of communication, which is used in a voluntary way and allows the rendering of what one wants to say.

This fundamental distinction between expressive acts and the language of mimicry carried the solution for a problem that confronted Darwin when he was searching for a criterion for authentic expression. Looking at the two photographs that Duchenne had made of the same old man: "Almost everyone recognized that the one represented a true, and the other a false smile; but I have found it very difficult to decide in what the whole amount of difference consists."[11] Later, Wittgenstein came back to this difficulty: "I may recognize a genuine loving look, distinguish it from a pretended one (and here there can, of course, be a 'ponderable' confirmation of my judgment). But I may be quite incapable of describing the difference."[12] Duchenne had suggested, though, that the criterion of authentic expression is not found where people usually look for it. The image of an authentic or simulated passion is indescribable, whatever the perspicacity of the observer. It is in relation to a type of organic functioning that it is possible to form the concept of the enunciation of authentic expressions. It suffices to describe the grammar of physiognomy in movement to take a normative posture. The difference between an authentic expression and a simulated one coincides with the difference between the muscular movements that express them. Once Duchenne identifies and describes the differential signs of authentic joy and simulated joy, the difficulty is overcome. "Is there such a thing as 'expert judgment' about the genuineness of expressions of feelings?" asked Wittgenstein. In the case of laughter, the answer is yes, specifying that this competence is acquired not by experience but by following the lessons of physiology. It would be wrong to underestimate the import of this line of separation that Duchenne was tracing between the false laugh and the honest laugh. By the false laugh, the person is actually saying something: sometimes contempt, "the laugh of Democritus" (Descartes), sometimes the joy of the character incarnated by the comedian, "the theatrical smile where one shows the teeth but the eyes do not smile" (Stendhal), and sometimes the smile of the person in society, "the smile of politeness" (Duchenne). By the frank laugh, a person expresses, without wanting to, something that he or she could never *say*. A joyous behavior without any joy—the case arising precisely from electrical simulation—is doubly impossible, because the person does not command the contraction of the inferior palpebral orbicular, and because he cannot say what is, by its very essence, *unsayable*. The springing forth of expressive force and its constraining character carry no symbolic determination. Only reflex actions realize the mechanism of the expressive act. Physiognomy presents an authentic expression when, and only when, the organism gives birth to the image of an emotion that it is impossible to mime.

Walter Benjamin said that in photography the exhibition value begins to suppress entirely the ritual value. But he noted that the latter does not cede ground without resisting: "It retires into an ultimate retrenchment: the human countenance. It is no accident that the portrait was the focal point of early photography. The cult of remembrance of loved ones, absent or dead, offers a last refuge for the cult value of the picture. For the last time the aura emanates from the early photographs in the fleeting expression of a human face."[13] Roland Barthes, although he compared photography to the theater instead, remained attached to the ritual value: "Photography is a kind of primitive theater, a kind of Tableau Vivant, a figuration of the motionless and made-up face beneath which we see the dead."[14] Duchenne's images relate less to the theater of actors than to the theater of marionettes. His theater of affects offers several points of contact with Henrich von Kleist's *Theater of Marionettes*. Duchenne's model is a "mannequin" as malleable as the marionettist's puppet; both may be led at the will of the showman. They relate to cords (strings and electrical wires) that are the means of animation. The mechanism of these figures relies on the same principle: each movement of the puppet has a center of gravity that it suffices to master inside the figure; similarly, each passion possesses a center of expression that it suffices to activate. Finally, the mechanical simplicity of both rhythmic and physiological movements presupposes the intervention of an expert machinist. Indifferent to the concern to "make it live," Duchenne, like the marionettist, circumvents the theme of the mythic denegation of the malaise of death that seduced Benjamin and Barthes so much. His images speak a stilted language that is as distant as possible from humanity and affectation. But photographs from the theater of affects show an expressive movement congealed in an impossible immobility. The acme of an expression expresses an exaggeration that may have two opposite effects: an effect of overflowing, because the medium accentuates the facial traits (hence the excess of rhetoric), and an effect of crushing, to the extent that the expression hardened by the medium loses its vivacity. To fix an expression that is only ever seen for an instant is to weight it with an indetermination that sometimes renders it unrecognizable. This conflict between the excess of rhetoric and the inevitable denaturation of the ephemeral fascinates the spectator, though. By its rigidity and its artificiality, the expression produces a *photogenic* effect. More than ever, the face is the site of the soul, but this site is deserted. Duchenne photographed it to find its indices. He was undoubtedly the first to give the photography of the human face *exhibition value.* That is why his photos arise from a study that is disquieting, surprising, and troubling for anyone who follows it. To penetrate them, one really feels that need of a direction, pathways, a caption.

With *Mécanisme de la physionomie humaine,* Duchenne put in place a structure of experience that has dominated the whole twentieth century. Not without methodological modifications, a series of interlocking research studies would be contained in it. All the authors of treatises on artistic anatomy, from Mathias Duval to Moreau via Paul Richer, Jean-Paul Morat, and Maurice Doyon, have merely taken up the descriptions and iconography of his book. The physiologists have continued to describe the various expressions in their relation to the diminution or augmentation of the neuromotor system. In the decade 1960 to 1970, Harold Scholsberger and Paul Ekman proceeded to an analysis of the forms and functions of the facial musculature that explores the terrain staked out by Duchenne. Psychologists still use the techniques and procedures that he invented to describe the forms of perception of facial expression and the innate mechanisms for treating information. They also borrow from him the principle of the synthetic construction of facial expressions based on minimal indicators. In their studies of behavior, Nico Fridja, Konrad Lorenz, and Irenäus Eibt-Eibesfeldt have reactivated, against the culturalists, the theme of an innate foundation for facial expressions. They have thus rediscovered the universal character of the expressive function already discovered by Duchenne. Theoreticians of dramatic acting, like Constantin Stanislavski and Elia Kazan, have also revived his intuitions about the force of expression in imaginary situations. It is not by chance that across the Atlantic Duchenne's works have retained the attention of specialists in plastic surgery. Finally, at the origin of modern techniques of imagery, which have displaced from the surface toward the brain the cartography of the passions, we find artisanal techniques of electrography and photography. It is rare for an author to open up so many new avenues. Duchenne of Boulogne, son of a corsair, was one of the first adventurers who went in conquest of surfaces. One might say of him what Nietzsche says of those who know how to pioneer a direction: "No river is great and bounteous through itself alone, but rather because it takes up so many tributaries and carries them onwards: that makes it great. It is the same with all great minds. All that matters is that one man give the direction, which the many tributaries must then follow."[15]

REFERENCE MATTER

Appendix

EXPRESSIVE MUSCLES CORRESPONDING TO PLATE 2,
ANATOMICAL PREPARATION OF THE MUSCLES OF THE FACE

Plate label	Term used in the text	French	Function
1.	Frontal	Frontal	Muscle of *attention*
2.	Superior orbicular palpebral	Orbiculaire palpébral supérieur	Muscle of *reflection*
3.	Superior and inferior palpebrals	Palpébraux supérieur et inférieur	Muscle of *scorn* and complementary to *crying*
4.	Inferior orbicular palpebral	Orbiculaire palpébral inférieur	Muscle of *benevolence* and complementary to *overt joy*
5.	Superciliary	Sourcilier	Muscle of *pain*
6.	Pyramidal of the nose	Pyramidal du nez	Muscle of *aggression*
7.	Zygomatic major	Grand zygomatique	Muscle of *joy*
8.	Zygomatic minor	Petit zygomatique	Muscle of *moderate crying* and of affliction
9.	Proper elevator of the upper lip	Élévateur propre de la lèvre supérieur	Muscle of *crying*
10.	Dilator of the nostrils	Dilatateur des ailes du nez	Complementary muscle to passionate expressions
n/a	Common elevator of the upper lip and nostril	Élévateur commun de la lèvre supérieure et de l'aile du nez	Muscle of *sniveling*
n/a	Masseter	Masséter	
n/a	Orbicular of the lips	Orbiculaire des lèvres	
12.	Deep fibers of the orbicular of the lips, continuous with the buccinator	Fibres profondes de l'orbiculaire des lèvres se continuant avec le buccinateur	
13.	Triangle of the lips	Triangulaire des lèvres	Muscle of *sadness* and complementary to *aggressive passions*
14.	Buccinator	Buccinateur	Muscle of *irony*

Plate label	Term used in the text	French	Function
15.	Square of the chin	Carré du menton	Complementary muscle of *irony* and *aggressive passions*
16.	Cleft of the chin	Houppe du menton	
17.	Transversal of the nose	Transverse du nez	Muscle of *lusciousness*, of *lubricity*
n/a	Muscle of the neck	Peaucier	Muscle of *fear*, of *terror*, and complementary to *anger*

SOURCE: From G.-B. Duchenne, *Mécanisme de la physionomie humaine ou analyse électro-physiologique de l'expression des passions* (Paris: Renouard, 1862).

Notes

FOREWORD

1. G.-B. Duchenne, *The Mechanism of Human Facial Expression*, ed. and trans. R. A. Cuthbertson (New York: Cambridge University Press, 1990; originally published 1862), part 1, "The Mechanism of Human Facial Expression or an Electrophysiological Analysis of the Expression of Emotions," p. 2. Duchenne had also published several papers on the subject before 1862, but his most noted work appeared in that year.

2. R. Descartes, *The Passions of the Soul*, trans. S. H. Voss (Indianapolis: Hackett, 1989; originally published in 1649). Ch. Le Brun, "Lecture on Expression: Conférence sur l'expression générale et particuliére," in *The Expression of the Passions*, ed. and trans. J. Montagu (New Haven and London: Yale University Press, 1994; originally published in 1667).

3. Ch. Darwin, *The Expression of the Emotions in Man and Animals*, 3rd ed. (New York: HarperCollins, 1998; originally published in 1872).

4. E. Levinas, "Ethics of the Face," in *Totality and Infinity: An Essay on Exteriority*, trans. A. Lingis (Pittsburgh: Duquesne University Press, 1969), p. 197.
The face resists possession, resists my powers. In its epiphany, in expression, the sensible, still graspable, turns into total resistance to the grasp. This mutation can occur only by opening of a new dimension. . . . The expression the face introduces into the world does not defy the feebleness of my powers, but my ability for power [*mon pouvoir de pouvoir*]. The face, still a thing among things, breaks through the form that nevertheless delimits it. This means concretely: the face speaks to me and thereby invites me to a relation incommensurate with a power exercised, be it enjoyment or knowledge (pp. 197–198).

5. Darwin, *The Expression of the Emotions*. For the epistemology of reflexive physiology, also see G. Canguilhem, *La formation du concept de réflexe aux XVII et XVIII siècles* (Paris: Presses Universitaires de France, 1955).

6. F. Delaporte, *Anatomie des passions* (Paris: Presses Universitaires de France, 2003), p. 101. All translations in the foreword are my own.

7. Duchenne, *Mechanism*, quoting Francis Bacon, *The New Organon*, ed. L. Jardine (New York: Cambridge University Press, 2000; originally published in 1620).

8. Duchenne, *Mechanism*, quoting George-Louis Leclerc Buffon, *Histoire naturelle* (Paris: Gallimard-Jeunesse, 1984; originally published in 1749).

9. Delaporte points out that electric stimulation had also been used in the attempt to reanimate dead flesh, and later would be used therapeutically in the treatment of psychiatric patients.

10. F. Delaporte, *La maladie de Chagas: Histoire d'un fleau continental* (Paris: Payot, 1999); *The History of Yellow Fever*, trans. A. Goldhammer (Cambridge, Mass.: MIT Press, 1991); *Disease and Civilization*, trans. A. Goldhammer (Cambridge, Mass.: MIT Press, 1989). See also F. Delaporte and P. Pinell, *Histoire des myopathies* (Paris: Payot, 1998).

11. Delaporte, *Anatomie des passions*, p. 199.

INTRODUCTION

1. J. Cruveilhier, *Traité d'anatomie descriptive*, 3rd ed. (Paris: Béchet et Labé, 1851), 3: 189.

2. F.-X. Bichat, "Essai sur Desault," in *Oeuvres chirurgicales de Desault* 1: 11; quoted by Michel Foucault in *The Birth of the Clinic*, trans. A. M. Sheridan Smith (New York: Vintage Books, 1973), p. 166.

3. G.-B. Duchenne, "Considérations générales sur la mécanique de la physionomie," Archives Nationales, Paris, 11 March 1857, F/17/3100/1, p. 28.

4. F. Nietzsche, *On the Genealogy of Morality*, trans. C. Diethe (New York: Cambridge University Press, 1994), p. 56.

5. G. Bachelard, *Le matérialisme rationnel* (Paris, 1953), p. 209.

6. G. L. L. Buffon, "De l'homme," in *Oeuvres de Buffon*, ed. Le Vasseur (Paris, 1884), 9: 61–62.

7. Duchenne's book is translated into English as *The Mechanism of Human Facial Expression*.

CHAPTER I

1. M. Bérard, "Application de la galvanisation localisée à l'étude des fonctions musculaires," *Bulletin de l'Académie nationale de médecine* 16 (1851), pp. 5–11.

2. G.-B. Duchenne, "Exposition d'une nouvelle méthode de galvanisation, dite galvanisation localisée," *Archives générales de médecine* 23 (1850), pp. 276–278.

3. J. J. Sue, *Élémens d'anatomie, à l'usage des peintres, des sculpteurs, et des amateurs*, 2nd ed. (Paris, 1788), p. iv.

4. P. Camper, *Discours prononcé par feu M. Pierre Camper, en l'académie de dessin d'Amsterdam, sur le moyen de représenter d'une manière sûre les diverses passions qui se manifestent sur le visage* (Utrecht: B. Wild et J. Altheer, 1792), p. 12.

5. J.-B. Sarlandière, *Physiologie de l'action musculaire appliquée aux arts d'imitation* (Paris: De Lachevardière, 1830), p. 7.

6. Cruveilhier, *Traité d'anatomie descriptive* (3rd ed), 3: 190.

7. J.-B. Winslow, *Exposition anatomique de la structure du corps humain* (Paris: G. Desprez and J. Desseartz, 1732), pp. 671a–672b. Some years earlier, J. Parsons had published "Human Physiognomy Explain'd: In the Crounian Lectures on Muscular Motion. For the year 1746. Read before the Royal Society." Supplement to the *Philosophical Transactions* 44 (1747), pp. 1–77. But Parsons said the same thing as Winslow: the frontal muscles determined the frontal lines and the gathering of the eyebrows.

8. J.-L. Moreau, "Analyse anatomique et physiologique du visage," in *L'Art de connaître les hommes par la physionomie*, 2nd ed. (Paris, 1820), 4: 220.

9. J. Cruveilhier, *Traité d'anatomie descriptive*, 2nd ed. (Paris: Béchet et Labé, 1843), 2: 197–198.

10. A. Béclard, "Face," *Dictionnaire de médecine*, 2nd ed. (Paris, 1835), 12: 322. See also H. Cloquet, *Traité d'anatomie descriptive*, 3rd ed. (Paris: Crochard, 1824), p. 440.

11. F.-X. Bichat, *Anatomie générale appliquée à la physiologie et à la médecine* (Paris: Brosson et Gabon, 1812), 3: 318–319.

12. G.-B. Duchenne, *De l'électrisation localisée et de son application à la physiologie, à la pathologie et à la thérapeutique* (Paris: J.-B. Baillière, 1855), p. 375.

13. Winslow, *Exposition anatomique*, p. 166a.

14. P. J. Barthez, *Nouveaux éléments de la science de l'homme* (Montpellier, 1778), p. 146.

15. Duchenne, "Considérations générales," p. 3.

16. Duchenne, *De l'électrisation localisée et de son application*, p. 277.

17. G. Canguilhem, "L'homme de Vésale dans le monde de Copernic: 1543," in *Études d'histoire et de philosophie des sciences* (Paris: Vrin, 1968), p. 33.

18. G.-B. Duchenne, *Physiologie des mouvements démontrés à l'aide de l'expérimentation électrique et de l'observation clinique et applicable à l'étude des paralysies et des déformations* (Paris: J.-B. Baillière, 1867), pp. v–vi.

19. G.-B. Duchenne, "Première note sur les fonctions des muscles de la face étudiées à l'aide de la galvanisation localisée," Archives de l'Académie des sciences, Paris, 12 March 1850, p. 5.

20. Duchenne, "Considérations générales," p. 4.

21. Duchenne, "Scientific Section," in *The Mechanism of Human Facial Expression*, p. 43. Duchenne mentioned the excellent experimental condition of his subject several times: "His face was insensitive, which allowed me to study the individual action of muscles with as much effectiveness as on a corpse. . . . My old man was thus a fitting subject for the demonstration of the physiological facts I needed to establish" (ibid., "Aesthetic Section," pp. 101–102).

22. Duchenne, *De l'électrisation localisée et de son application*, 3rd ed., p. 1023n1. J. T. Hueston and R. A. Cuthbertson are wrong to write that "in 1850 Duchenne had given the original ensemble of his research on the muscles of the face" ("Duchenne de Boulogne and Facial Expression," *Annals of Plastic Surgery* 1 (1978), p. 441).

23. G.-B. Duchenne, "Troisième et dernière note sur les fonctions des muscles de la face étudiées à l'aide de la galvanisation localisée," Archives de l'Académie des sciences, Paris, 12 March 1850, p. 15. Some lines later: "Moreover, did there not exist an anatomical reason for us to admit with Albinus that the superciliary is a root of the orbicular? I am speaking of the intimate union of its fibers with those of the orbicular in the skin of the eyebrow." In the first edition of *De l'électrisation localisée et de son application* (1855), Duchenne gives nothing more than a résumé of his manuscript papers of 1850; see chapter 4, "Électro-physiologie des muscles de la face," pp. 373–392.

24. Duchenne, "Recherches anatomiques et expérimentales sur les muscles du sourcil," in *Physiologie des mouvements*, p. 827.

25. Cruveilhier, *Traité d'anatomie descriptive*, 2nd ed., 2: 34n1.

CHAPTER 2

1. Darwin, introduction to *The Expression of the Emotions in Man and Animals*, pp. 7–21.

2. R. Descartes, article 113, "The Passions of the Soul," in *The Philosophical Works*, trans. E. Haldane and G. R. T. Ross (New York: Cambridge University Press, 1968), 1: 381.

3. R. Descartes, "A Discourse on Method," part 5, p. 117. See also "Letter to the Marquis of Newcastle, 23 November 1646": "For the movements of our passions, although they are accompanied in us by thought, because we have the faculty of thinking, it is nevertheless very evident that they do not depend on it."

4. M. Cureau de La Chambre, *L'Art de connaître les hommes* (Paris: P. Rocolet, 1659), p. 2.

5. Ibid., p. 112.

6. G. Lavater, "Rapports de la physiognomonie avec la peinture," in *L'Art de connaître les hommes par la physionomie*, 2nd ed. (Paris, 1820), 8: 11.

7. G. Lavater, "Quelques vues générales sur la physiognomonie," in *L'Art de connaître les hommes par la physionomie* 3: 205.

8. G. Lavater, "Introduction et considérations générales," in *L'Art de connaître les hommes par la physionomie* 1: 226.

9. G. Lavater, "De l'art de voir et d'observer les physionomies," in *L'Art de connaître les hommes par la physionomie* 5: 24–25.

10. Winslow, *Exposition anatomique*, p. 725b.

11. J.-L. Moreau, "Observations physiologiques sur l'expression et les caractères des passions," in *L'Art de connaître les hommes par la physionomie* 5: 221–222.

12. Moreau, "Analyse anatomique et physiologique du visage," 4: 224–225.

13. Ibid., p. 207.

14. Duchenne, *Mechanism*, p. 5.

15. Moreau, "Analyse anatomique et physiologique du visage," p. 212.

16. E. Lacan, *Esquisses photographiques à propos de l'exposition universelle* (Paris: Grassart, 1856), pp. 130–131.

17. S. Bula and M. Quétin, "Duchenne de Boulogne et le Prix Volta," in *Duchenne de Boulogne* (Paris: École nationale supérieur des beaux-arts, 1999), pp. 55–56. "These various marks left by Adrien Tournachon ought to lead to prudence by specialists today who think they should minimize his importance. From his art emanates a skittish temper and a characteristic weight that could only be from him." J.-F. Debord, "Une leçon de Duchenne," in *Duchenne de Boulogne* (Paris: École nationale supérieur des beaux-arts, 1999), p. 32n20.

18. F. Nadar, *Revendication de la propriété exclusive de pseudonyme Nadar* (Paris, 1857), quoted by M. Frizot and F. Ducros, *Du bon usage de la photographie: Une anthologie de textes choisis* (Paris: Centre National de la Photographie, 1987), p. 9.

19. Duchenne, "Foreword to 'Scientific Section,'" in *Mechanism*, p. 39n.

20. A. Jammes, "Duchenne de Boulogne: La grimace provoquée et Nadar," *Gazette des Beaux-Arts*, 92, no. 1319 (1978), p. 217a. See also A. Jammes and R. Sobieszek: "Adrien Tournachon (Nadar jeune) and his brother Félix, the famous Nadar, combined their efforts. From the start, they were capable of analyzing scientifically the most detailed expression of the human face"; in "Commentary on photograph 197, plate 72," *French Primitive Photography* (Philadelphia: Philadelphia Museum of Art, 1969).

21. F. Heilbrun, "L'Art du portrait photographique chez Félix Nadar," in *Nadar, Les années créatices: 1854–1860* (Paris: Réunion des musées nationaux, 1994), pp. 58–59. But if Nadar had collaborated with his brother Adrien on the photos of the mime, he would not have written, "I would not even hesitate to assert that nobody has seen anything superior to the great heads of impression [*sic*] of the mime Deburau *fils* by Adrien Tournachon (another escapee from painting)"; Nadar, *Quand j'étais photographe* (Arles: Acte Sud, 1998), p. 103.

22. S. Aubenas, "Au-delà du portrait, au-delà de l'artiste," in *Nadar, Les années créatices: 1854–1860* (Paris: Réunion des musées nationaux, 1994), p. 162.

23. Ch. Hacks, *Le Geste* (Paris: Marpon et Flammarion, 1892), p. 339.

24. Duchenne, *Mechanism*, p. 43.

25. Ibid., p. 39.

26. Ibid., p. 36.

27. Duchenne, "Considérations générales," p. 4.

28. G.-B. Duchenne, Preface to *Album de photographies pathologiques* (Paris: J.-B. Baillière, 1862), unpaginated.

29. Duchenne, *De l'électrisation localisée*, 3rd ed., p. 1023n1, my emphasis. The photographs of experiments were made *after 1855*, at the Charity Hospital in the clinic of M. Briquet, and not at the Salpêtrière Hospital (see above, p. 27 from chapter 1, handwritten version). Moreover, the date 1856 is confirmed by the only trace that can be found; cf. S. Aubenas, "Lettre du 25 juillet 1856 adressé par

Adrien Tournachon à Poitevin" (BN, Manuscript Dept., "Fonds Poitevin prints of Duchenne lithography," cited by F. Heilbrun, "L'Art du portrait photographique chez Félix Nadar," in *Nadar*, p. 97n87. And in the foreword to the "Aesthetic Section," Duchenne says that the photographs that compose the "Scientific Section" "date mostly from 1856" (p. 101).

30. Duchenne, "Recherches anatomiques et expérimentales sur les muscles du sourcil," in *Physiologie des mouvements*, p. 827.

31. Duchenne, "Considérations générales," p. 11. Duchenne gave a second version of his discovery. After an accidental fall of the veil at the moment of exciting the superciliary, he saw the immobility of the lower half of the face. This version is not credible. Either "it is impossible to not let oneself be fooled by this illusion" and there is no reason to imagine challenging it, or else the *conditions of expression* (electrization of the mortuary mask) contradict the evidence of sense perception, and the latter is the problem (cf. Duchenne, "Principal Facts," in *Mechanism*, pp. 12–19).

32. Duchenne, "Principal Facts," p. 15.

33. Ibid., p. 17.

34. Diderot, "Essai sur la peinture," in *Oeuvres esthétiques*, ed. P. Vernière (Paris, 1994), pp. 712–715.

35. Duchenne, "Considérations générales," p. 24.

36. Duchenne, "The Purpose of My Research," in *Mechanism*, p. 34. This is another reason for Duchenne's interest in his elderly subject: "Due to the aging process he had developed all the lines produced by the expressive muscles, lines that I have divided into principal lines constituting the expression, and secondary lines indicating the age of the subject and the different degree of expressive movement" (foreword to the "Aesthetic Section," p. 101).

37. Ibid., p. 37.

38. "Discours de M. le Professeur Mathias Duval," in *Inauguration du monument élevé à la mémoire de Duchenne de Boulogne* (Paris, 1897), p. 28.

39. "Rapport sur l'ouvrage de M. le Docteur Duchenne de Boulogne," Minutes of the Commission Meetings, Archives Nationales, 1857, F/17/3100/1.

40. Darwin, *Expression of the Emotions in Man and Animals*, p. 21 (published in French in 1890). Verneuil had already insisted on this point: "To well understand the whole work and its originality, it is necessary to have the plates before one's eyes while reading the text"; review of Duchenne's book in *Gazette hebdomadaire de médecine et de chirurgie* (1862), 28: 448a.

41. Duchenne, *Mechanism*, pp. 90–92.

CHAPTER 3

1. Rapport de M. Serres sur l'ouvrage de M. Duchenne de Boulogne, intitulé *Mécanisme de la physionomie humaine*, Archives nationales, meeting 26 May 1864, F/17/3101, pp. 1–15.

2. Duchenne, *Mechanism*, p. 12. By "partial contraction" Duchenne means a simple or isolated contraction of a muscle, in contrast to synergetic or composite contractions. Cf. Duchenne, "Considérations générales," p. 10b.

3. Duchenne, *Mécanisme de la physionomie humaine*, pp. 52–53. G. Dumas and A. Jammes have therefore made the same mistake in reading by assimilating provoked expressions to grimaces. Cf. G. Dumas, "L'Expression des émotions," in *Traité de psychologie* (Paris, 1923), 2: 632; A. Jammes, "Duchenne de Boulogne," pp. 215–220.

4. Duchenne, *Mechanism*, p. 72.

5. Duchenne, *Mécanisme de la physionomie humaine*, pp. 69–70.

6. Ibid., preface, pp. xi–xii.

7. Ibid., p. 26.

8. Duchenne, "Considérations générales," p. 29.

9. Ibid.

10. Duchenne, *Mécanisme de la physionomie humaine*, pp. 47–49.

11. Cl. Bernard, *De la physiologie générale* (Paris: Hachette, 1872), p. 254n59.

12. P. Gratiolet, *De la physionomie et des mouvements d'expression* (Paris: J. Hetzel, 1865), pp. 8–9.

13. E. Brissaud, "L'Oeuvre scientifique de Duchenne de Boulogne," *Revue internationale d'électrothérapie et de radiologie* 3 (1899), p. 81.

14. A. Lemoine, *L'Âme et le corps*, 1862, p. 1. See also V. Egger, "La physiologie cérébrale et la psychologie," *Revue des deux mondes* 24 (1877), p. 209: "The science of organic movements will not become the science of immaterial facts; it will not absorb psychology."

15. A. Dechambre, "Anatomie des beaux-arts," *Dictionnaire encyclopédique des sciences médicales* (1867), 4: 247–249. Cuyer's criticisms are of the same kind; cf. *La mimique* (Paris: Octave Doin, 1902), pp. 41ff.

16. Duchenne, "Considérations générales," p. 16n1.

17. Gratiolet, *De la physionomie et des mouvements d'expression*, p. 8.

18. Dechambre, "Anatomie des beaux-arts," p. 255.

19. Brissaud, "L'Oeuvre scientifique de Duchenne de Boulogne," p. 81.

20. "Rapport de M. Serres," p. 7.

21. Darwin, *Expression of the Emotions in Man and Animals*, p. 18, quoting Duchenne, *Mechanism*, p. 19.

22. Ch. Darwin, *The Origin of Species* (New York: Modern Library, 1998), pp. 579–581. This critique is taken up by Th. Ribot, *La psychologie des sentiments* (Paris: Presses Universitaires de France, 1939), 16th ed., p. 127n2, and by P. Guilly, *Duchenne de Boulogne* (Paris: J.-B. Baillière et fils, 1936), p. 211.

23. G.-B. Duchenne, "Recherches électro-physiologiques, pathologiques et thérapeutiques sur le diaphragme," *L'Union médicale* 101 (1853), p. 26.

24. Darwin, *Expression*, p. 132.

25. P. Gratiolet, Conférence by M. Gratiolet (soirées scientifiques de la Sorbonne), ed. E. Algrave, "De l'homme et de sa place dans la création," *Revue des cours scientifiques* 16 (19 March 1864), p. 193b.

26. Cruveilhier, *Traité d'anatomie descriptive*, 3rd ed., 3: 209–211.

27. Ibid., edition revised and corrected by Marc Sée and Cruveilhier's son, 4th ed. (Paris: P. Asselin, 1862–1867), 1: 608–615.

28. Verneuil, "Mécanisme de la physionomie humaine," pp. 445a–b.

29. Duchenne, *Mechanism*, p. 29.

30. E. B. de Condillac, *Principes généraux de grammaire*, new ed. (Paris: A. J. Ducour), p. 9. See also *Essai sur l'origine des connaissances humaines* (Paris: Alive, 1998; 1746), pp. 164–165.

31. J.-J. Virey, "Physionomonie," *Dictionnaire des sciences médicales* (Paris: Panckoucke, 1820), 42: 216.

32. Duchenne, *Mechanism*, p. 30.

33. Descartes, *The Passions of the Soul*, trans. S. H. Voss, p. 86.

34. Duchenne, *Mechanism*, pp. 29–30.

35. *Mécanisme de la physionomie humaine*, "Partie esthétique," p. 129.

36. F. Nietzsche *The Twilight of the Gods* (New York: Oxford University Press, 1998). See Michel Foucault's commentary: "God is perhaps less beyond knowledge than a certain distance on this side of our phrases; and if Western man is inseparable from him, it is not by an invincible propensity to break through frontiers of experience, but because his language unceasingly foments him in the shadow of his law"; *The Order of Things: An Archaeology of Human Sciences* (New York: Vintage Books, 1994), p. 54.

37. D. Diderot, "Notes on Painting" (1768), in *Diderot on Art*, trans. J. Goodman (New Haven, Conn.: Yale University Press, 1995), pp. 210–212.

38. Duchenne, *Mechanism*, p. 31. What the movements of a good soul could not modify was the alteration of facial features due to local afflictions of the face (spasms, partial paralyses, tics).

39. Verneuil, "Mécanisme de la physionomie humaine," p. 447b. Claude Bernard compares the mechanism of expression to "a musical instrument that can render melodies of which it is totally unaware"; *Physiologie générale*, p. 254n59. And Guyot writes: "False notes will not be any more tolerated in the plastic arts than in music"; "Psychologie: L'art et la science," *Revue scientifique* 14, 3rd ser. (1887), pp. 144a–b.

40. Descartes, *The Passions of the Soul*, trans. S. H. Voss, p. 101.

41. D. Diderot, "Paradox in Acting," in *Diderot's Selected Writings*, ed. L. Crocker (New York: Macmillan, 1966), p. 323.

42. Duchenne, *Mechanism*, p. 30.

43. G. Deleuze, *Logic of Sensation*, trans. D. W. Smith (Minneapolis: University of Minnesota Press, 2003).

CHAPTER 4

1. A. Latour, "Mécanisme de la physiologie humaine," *L'Union médicale* 103 (1862), pp. 422–423.

2. R. de Piles, *Cours de peinture par principes* (Nîmes: J. Chambon, 1990), p. 29. For Duchenne's answer to Latour's criticism, see "Aesthetic Section," in *Mechanism*, pp. 109–110n1.

3. R. W. Lee, *Ut Pictura Poesis: The Humanistic Theory of Painting* (New York: Norton, 1967), p. 27. See also N. Bryson, *Word and Image: French Painting of the Ancient Regime* (New York: Cambridge University Press, 1981).

4. J. Montagu, *The Expression of the Passions* (New Haven and London: Yale University Press, 1994), p. 17. On Nivelon, cf. ms. 12987 of the French Collection of the Bibliothéque Nationale, p. 211; quoted by A. Fontaine, *Les doctrines d'art en France* (Paris, 1909), p. 101.

5. Descartes, *The Passions of the Soul*, trans. S. H. Voss, art. 27, pp. 33–34.

6. Le Brun, "Conférence sur l'expression générale et particulière," in *L'Art de connaître les hommes par la physiognomonie*, 2nd ed. (Paris: L.-T. Cellot, 1820), 9: 268–269. Moreau reproduces Picard's edition (1698). Le Brun writes: "Today I shall try to demonstrate that expression is also that which reflects the movements of the soul, which makes visible the effects of passions" (p. 262).

7. Ibid.; Le Brun, "Sur le Saint Michel terrassant le démon de Raphaël" (7 May 1667), in *Les Conférences de l'Académie royale de peinture et de sculpture aux XVIII siècle*, ed. A. Mérot (Paris: École nationale supérieur des Beaux-Arts, 1996), p. 63.

8. J.-B. Du Bos, *Réflexions critiques sur la poésie et sur la peinture* (Paris: École nationale supérieure des Beaux-Arts, 1993), pp. 133–134.

9. Le Brun, "Conférence sur l'expression," p. 261.

10. Ibid.; "Sur Les Israelites receuillant la manne dans le désert de Poussin" (5 November 1667), in *Les Conférences de l'Académie royale*, p. 111.

11. Cureau de La Chambre, *Les caractères des passions* (Paris: P. Rocolet, 1640), 1: 19. On mind mechanisms, see *L'Art de connaître les hommes*, pp. 141ff.

12. E. H. Gombrich, *Art and Illusion: A Study in the Psychology of Pictorial Representation* (New York: Pantheon, 1960), p. 348. See also D. Posner, "Charles Le Brun's *Triumphs of Alexander*," *Art Bulletin* 91 (1959), pp. 237–247.

13. H. Damisch, "L'Alphabet des masques," *Nouvelle revue de psychoanlyse* 21 (1980), p. 130.

14. E. Pommier, *Théories du portrait: De la Renaissance aux Lumières* (Paris: Gallimard, 1998), p. 213.

15. Montagu, *The Expression of the Passions*, pp. 90–91.

16. P. Camper, *Discours prononcé par feu M. Pierre Camper*, p. 6. Cf. W. Hogarth, *Analyse de la beauté* (Paris: École nationale supérieur des Beaux-Arts, 1991), p. 158. M.-F. Dandré-Bardon wrote: "We refer the curious to this masterpiece of

pictorial literature if they want to instruct themselves in the most circumstantial details of which expression is susceptible"; *Traité de peinture suivi d'un essai sur la sculpture* (Paris, 1765), p. 78.

17. Comte A.-C. Caylus, "Papier du Comte de Caylus: De l'étude de la tête en particulier," Archives de l'École de Beaux-Arts, 1: 69–71. Cf. M. Duvivier, "Liste des peintres et des sculpteurs couronnés jusqu'en 1861 dans le concours de la tête d'expression," *Archives de l'art français* (Paris, 1861), 1: 195–208.

18. E. J. Delécluze, *Louis David, son école et son temps* (Paris: Macula, 1983), p. 227.

19. Letter from David during his exile, 7 November 1817, communicated by M. B. Fillon, *Nouvelles archives de l'art francais*, ser. 1 (1875), p. 427. Cf. the article by D. Johnson, "Desire Demythologised: David's *L'Amour quittant Psyché*," *Art History* 4 (1986), pp. 450–470.

20. J. J. Winckelmann, *Reflections on the Imitation of Greek Works in Painting and Sculpture*, trans. E. Heyer and R. C. Norton (La Salle, Ill.: Open Court, 1987), p. 33. (Editor's note: The translation has been modified with insights from Henry Fuseli's 1765 edition, p. 30.)

21. J. J. Winckelmann, *History of the Art of Antiquity*, trans. H. F. Mallgrave (Los Angeles: Getty Research Institute, 2006), part 1, p. 207.

22. Ibid., 2: 313.

23. G. E. Lessing, *Laocoön: An Essay on the Limits of Painting and Poetry*, trans. E. A. McCormick (Baltimore: Johns Hopkins University Press, 1984), p. 17.

24. J. W. Goethe, "On the *Laocoön* Group" (1798), in *Essays on Art and Literature*, ed. J. Gearey, trans. E. von Nardroff and E. H. von Nardroff (New York: Suhrkamp, 1986), p. 15.

25. Ch. Bell, *Essays on the Anatomy and Philosophy of Expression*, 2nd ed. (London: John Murray, 1824), p. 135.

26. J. W. Goethe, "L'Essai sur la peinture de Diderot" (1799), in *Écrits sur l'art* (Paris: GF-Flammarion, 1996), p. 166.

27. Duval, "Discours de M. le Professeur Mathias Duval," p. 30. On this celebration of the union of science and art, see A. Couder, "De l'application des recherches de M. Duchenne de Boulogne aux arts plastiques," *Journal des débats*, 22 septembre 1863, and Y. Guyot, "Psychologie: L'art et la science," *Revue scientifique*, 1887, ser. 3, 14: 142b.

28. P. Richer, *Artistic Anatomy*, ed. and trans. R. B. Hale (New York: Watson-Guptill, 1971).

29. Dechambre, "Anatomie des beaux-arts," p. 235.

30. Latour, "Mécanisme de la physiologie humaine," p. 422.

31. Diderot, "Notes on Painting."

32. Ch. Baudelaire, "The Modern Public and Photography," in *Art in Paris, 1845–1862*, ed. and trans. J. Mayne (London: Phaidon, 1965), p. 152.

33. Duchenne, *Mechanism*, pp. 35–36.

34. Ibid., p. 101.

35. Ibid., p. 110n1.

36. G.-B. Duchenne, "Étude physiologique sur la courbure lombo-sacrée et sur l'inclinaison du bassin pendant la station droite." Reprint. *Archives générales de médecine* 8 (1866), p. 16.

37. A. Lemoine, *La physionomie et la parole* (Paris: Gemer Baillière, 1865), p. 80.

38. C-H. Watelet and M. Lévesque, "Détail," *Dictionnaire des Arts de peinture, sculpture, et gravure* (Paris, 1792), 1: 618. Cf. D. Arasse, *Le Détail* (Paris: Flammarion, 1992), pp. 40–42.

39. M. Foucault, "Photogenic Painting," trans. P. A. Walker, *Critical Texts* 6, no. 3 (1989), pp. 1–2.

40. G. Courbet, "Le précurseur d'Anvers du 22 aout 1861," in *Peut-on enseigner l'art?* (Caen: L'Échoppe, 1986), unpaginated.

41. A. Disdéri, *L'Art de la photographie* (Paris, 1862); Frizot and Ducros, *Du bon usage de la photographie*, p. 46.

42. Duchenne, *Mechanism*, p. 115.

43. Ibid., p. 118.

44. Disdéri, *L'Art de la photographie*, p. 45.

45. Duchenne, *Mechanism*, p. 111.

46. Ibid., p. 110.

47. J. Ruskin, *Lectures on Architecture and Painting, Delivered at Edinburgh in November 1853* (New York: Charles E. Merrill, 1892), p. 218, paragraph 131.

48. Baudelaire, "The Modern Public and Photography, Salon of 1859," p. 155.

49. Ibid., p. 153.

50. Duchenne, *Mechanism*, pp. 111–112. To this series should be added figure 84: *Lady Macbeth receives King Duncan with a perfidious smile*. In the same spirit, see also the series corresponding to figures 75 to 78, *A Nun Pronouncing Her Vows*.

CONCLUSION

1. A. R. Damasio, *Descartes' Error: Emotion, Reason, and the Human Brain* (New York: Grosset Putnam, 1994), pp. 141–142.

2. J.-J. Courtine and C. Haroche, *Histoire du visage* (Paris: Payot, 1994), pp. 265–266.

3. R. Sobieszek, *Ghost in the Shell* (Cambridge, Mass., and London: MIT Press, 2000), p. 76.

4. Quoted by Guilly, *Duchenne de Boulogne*, p. 212.

5. L. Wittgenstein, *Philosophical Investigations*, trans. G. E. M. Anscombe (New York: Macmillan, 1953), p. 53.

6. P. Rayer, "Rapport sur l'ouvrage de M. le Docteur Duchenne de Boulogne."

7. A.-M. Hans-Drouin, *La communication non verbale avant la lettre* (Paris: L'Harmattan, 1995), p. 91.

8. Duchenne, "Scientific Section," in *Mechanism*, p. 44.

9. Duchenne, "Considérations générales," p. 2.

10. Duchenne, "Principal Facts," in *Mechanism*, pp. 14–15.

11. Darwin, *Expression of the Emotions in Man and Animals*, pp. 354–355.

12. Wittgenstein, *Philosophical Investigations*, p. 228.

13. W. Benjamin, "The Work of Art in the Age of Mechanical Reproduction," in *Illuminations: Essays and Reflections*, ed. H. Arendt, trans. H. Zorn (New York: Schocken, 1969), pp. 225–226.

14. R. Barthes, *Camera Lucida: Reflections on Photography*, trans. R. Howard (New York: Hill and Wang, 1981), p. 32.

15. F. Nietzsche, *Human, All Too Human: A Book for Free Spirits*, trans. M. Faber (Lincoln: University of Nebraska Press, 1984), p. 239.

Bibliography

Arasse, D. *Le détail.* Paris: Flammarion, 1992.

Archer, F. S. *Manual of the Collodion Photographic Process.* London: Printed for the Author, 1852.

Aubenas, S. "Au-delà du portrait, au-delà du portrait, au-delà de l'artiste." In *Nadar: Les années créatrices, 1854–1860.* Paris: Réunion des musées nationaux, 1994, pp. 152–167.

Bachelard, G. *Le matérialisme rationnel.* Paris: PUF, 1953.

Barthes, R. *Camera Lucida: Reflections on Photography.* Translated by R. Howard. New York: Hill and Wang, 1981.

———. *La chambre claire.* Paris: Gallimard Seuil, 1980.

Barthez, P. J. *Nouveaux éléments de la science de l'homme.* Montpellier, 1778.

Baudelaire, Ch. "The Modern Public and Photography." Edited and translated by J. Mayne. In *Art in Paris, 1845–1862.* London: Phaidon, 1965.

———. "Le public moderne et la photographie" (Salon de 1859). In *Écrits esthétiques.* Paris, 10/18, 1986, pp. 285–291.

Béclard, A. "Face." *Dictionnaire de médecine.* 2nd ed. Paris, 1835, vol. 12, pp. 521–525.

Bell, Ch. *Essays on the Anatomy and Philosophy of Expression.* 2nd ed. London: John Murray, 1824.

———. *Exposition du système naturel des nerfs du corps humain, suivie des mémoire sur le même sujet, lus devant la Société Royale de Londres.* Translated by J. Genest. Paris: L.-H. Hérhan, 1825.

Benjamin, W. "L'Oeuvre d'art à l'époque de sa reproduction mécanisée" (1936). In *Écrits français,* Présentés par J.-M. Monnoyer. Paris: Gallimard, 1991, pp. 117–194.

———. "The Work of Art in the Age of Mechanical Reproduction." In *Illuminations: Essays and Reflections,* ed. H. Arendt, trans. H. Zorn. New York: Schocken, 1969.

Bérard, M. "Application de la galvanisation localisée à l'étude des fonctions musculaires." *Bulletin de l'Académie nationale de médecine,* 1851, *16,* pp. 5–25.

Bernard, Cl. *De la physiologie générale.* Paris: Hachette, 1872.

———. *Rapport sur les progrès et la marche de la physiologie générale en France.* 1867.

Bichat, F.-X. *Anatomie générale appliquée à la physiologie et à la médecine.* Paris: Brosson et Gabon, 1812, vol. 3.

Brissaud, É. "L'Oeuvre scientifique de Duchenne de Boulogne." *Revue internationale d'électrothérapie et de radiologie,* 1899, *3,* pp. 69–92.

Bryson, N. *Word and Image: French Painting of the Ancient Regime.* New York: Cambridge University Press, 1981.

Buffon, Comte G. L. L. "De l'homme." In *Oeuvres de Buffon,* ed. Le Vasseur. 1884, vol. 9.

Bula, S., and M. Quétin. "Duchenne de Boulogne et le Prix Volta." In *Duchenne de Boulogne.* Paris: École nationale supérieur des beaux-arts, 1999, pp. 51–66.

Camper, P. *Discours prononcé par feu M. Pierre Camper, en l'académie de dessin d'Amsterdam, sur le moyen de représenter d'une manière sûre les diverses passions qui se manifestent sur le visage.* Utrecht: B. Wild et J. Altheer, 1792.

Canguilhem, G. *La formation du concept de réflexe aux XVII et XVIII siècles.* Paris: Presses Universitaires de France, 1955.

———. "L'Homme de Vésale dans le monde de Copernic: 1543." In *Études d'histoire et de philosophie des sciences.* Paris: Vrin, 1968, pp. 27–36.

Caylus, Comte A.-C. "Papier du Comte de Caylus: De l'étude de la tête en particulier." Archives de l'École des beaux-arts, ms. 522, vol. 1, ff. 69–73.

Chevreul, M.-E. *De la loi du contraste simultané des couleurs.* Paris: Pitois-Levrault, 1839.

Cloquet, H. *Traité d'anatomie descriptive.* 3rd ed. Paris: Crochard, 1824.

de Condillac, É. B. *Essai sur l'origine des connaissances humaines.* Paris: Alive, 1998 (1746).

———. *Principes généraux de grammaire.* Paris: A. J. Ducour, 1798.

Couder, A. "De l'application des recherches de M. Duchenne de Boulogne aux arts plastiques." *Journal des débats,* du 22 septembre 1863.

Courbet, G. "Le précurseur d'Anvers du 22 août 1861." In *Peut-on enseigner l'art?* Caen: L'Échoppe, 1986.

Courtine, J.-J., and C. Haroche. *Histoire du visage.* Paris: Payot, 1994 (Rivages, 1988).

Cruveilhier, J. *Traité d'anatomie descriptive.* 2nd ed. Paris: Béchet et Labé, 1843, vol. 2.

———. *Traité d'anatomie descriptive.* 3rd ed. Paris: Béchet et Labé, 1851, vol. 3.

———. *Traité d'anatomie descriptive.* Revue et corrigée par Marc Sée et Cruveilhier fils. 4th ed. Paris: P. Asselin, 1862–1867, vol. 1.

Cureau de La Chambre, M. *L'Art de connaître les hommes.* Paris: P. Rocolet, 1659.

———. *Les caractères des passions.* 4 vols. Paris: P. Rocolet, 1640–1662.

Cuthbertson, A., and J. T. Hueston. "Duchenne de Boulogne and Facial Expression." *Annals of Plastic Surgery*, 1978, *1*, pp. 411–420.

Cuyer, É. *La mimique*. Paris: Octave Doin, 1902.

Damasio, A. R. *Descartes' Error: Emotion, Reason, and the Human Brain*. New York: Grosset Putnam, 1994.

Damisch, H. "L'Alphabet des masques." *Nouvelle revue de psychanalyse*, 1980, *21*, pp. 123–131.

Dandré-Bardon, M.-F. *Traité de peinture suivi d'un essai sur la sculpture*. Paris, 1765.

Darwin, Ch. *L'Expression des émotions chez l'homme et les animaux*. 2nd French ed. Paris: C. Reinwald, 1890.

———. *The Expression of the Emotions in Man and Animals*. 3rd ed. New York: HarperCollins, 1998; originally published in 1872.

———. *De l'origine des espèces*. 3rd French ed. Paris: Guillaumin et Cie, Victor Masson, 1870.

———. *The Origin of Species*. New York: Modern Library, 1998.

David, J.-L. "Lettre de David pendant son exil, du 7 novembre 1817 communiquées par M. B. Fillon." *Nouvelles Archives de l'art français*, 1875, 1st series, pp. 425–429.

Debord, J.-F. "Une leçon de Duchenne." In *Duchenne de Boulogne*. Paris: École nationale supérieur des beaux-arts, 1999, pp. 27–40.

Dechambre, A. "Anatomie des beaux-arts." *Dictionnaire encyclopédique des sciences médicales*. 1867, vol. 4, pp. 231–250.

Delaporte, F. *Anatomie des passions*. Paris: Presses Universitaires de France, 2003.

———. *Disease and Civilization*. Translated by A. Goldhammer. Cambridge, Mass.: MIT Press, 1989.

———. *The History of Yellow Fever*. Translated by A. Goldhammer. Cambridge, Mass.: MIT Press, 1991.

———. *La maladie de Chagas: Histoire d'un fleau continental*. Paris: Payot, 1999.

Delaporte, F., and P. Pinell. *Histoire des myopathies*. Paris: Payot, 1998.

Delécluze, E. J. *Louis David, son école et son temps*. Paris: Macula, 1983 (1855).

Deleuze, G. *Logic of Sensation*. Translated by D. W. Smith. Minneapolis: University of Minnesota Press, 2003.

———. *Logique du sens*. Paris: Minuit, 1969.

Descartes, R. "Discours de la méthode." In *Oeuvres et lettres*, ed. Bridoux. Paris: Gallimard, 1953, pp. 126–179.

———. "Lettre au Marquis de Newcastle, 23 novembre 1646." In *Oeuvres et lettres*, pp. 1252–1256.

———. *The Passions of the Soul*. Translated by S. H. Voss. Indianapolis: Hackett, 1989; originally published in 1649.

————. *The Philosophical Works,* vol. 1. Translated by E. Haldane and G. R. T. Ross. New York: Cambridge University Press, 1968.

————. "Traité des passions." In *Oeuvres et lettres,* pp. 693–801.

Diderot, D. "Essai sur la peinture." In *Oeuvres esthétiques de Diderot,* ed. P. Vernière. Paris: Classiques Garnier, 1994, pp. 665–740.

————. "Notes on Painting" (1768). In *Diderot on Art,* trans. J. Goodman. New Haven, Conn.: Yale University Press, 1995.

————. "Paradox in Acting." In *Diderot's Selected Writings,* ed. Lester Crocker. New York: Macmillan, 1966.

————. "Le paradoxe sur le comédien." In *Oeuvres esthétiques de Diderot,* pp. 299–386.

————. "Pensées détachées sur la peinture." In *Oeuvres esthétiques de Diderot,* pp. 749–840.

Disdéri, A. *L'Art de la photographie.* Paris, 1862.

Du Bos J.-B. *Réflexions critiques sur la poésie et sur la peinture.* Paris: École nationale supérieur des beaux-arts, 1993 (1740).

Duchenne, G.-B. *Album de photographies pathologiques.* Paris: J.-B. Baillière, 1862.

————. "Considérations générales sur la mécanique de la physionomie." Archives Nationales, Paris, 11 March 1857, F/17/3100/1, ff. 1–33.

————. *De l'électrisation localisée et de son application à la physiologie, à la pathologie et à la thérapeutique.* Paris: J.-B. Baillière, 1855; 2nd ed., 1861; 3rd ed., 1872.

————. "Étude physiologique sur la courbure lombo-sacrée et sur l'inclinaison du bassin pendant la station droite." Reprint. *Archives générales de médecine,* 1866, *8,* pp. 1–16.

————. "Exposition d'un nouvelle méthode de galvanisation, dite galvanisation localisée." *Archives générales de médecine,* 1850, *23,* pp. 257–289.

————. *Mécanisme de la physionomie humaine ou analyse électro-physiologique de l'expression des passions.* Paris: Renouard, 1862.

————. *The Mechanism of Human Facial Expression.* Edited and translated by R. A. Cuthbertson. New York: Cambridge University Press, 1990 (originally published 1862).

————. "Note sur le spasme fonctionnel et la paralysie musculaire fonctionnelle." *Bulletin général de thérapeutique,* 1860, *58,* pp. 145–151, 196–202, 245–251.

————. *Physiologie des mouvements démontrés à l'aide de l'expérimentation électrique et de l'observation clinique et applicable à l'étude des paralysies et des déformations.* Paris: J.-B. Baillière, 1867.

————. "Première note sur les fonctions des muscles de la face étudiées à l'aide de la galvanisation localisée," Archives de l'Académie des sciences, Paris, 12 March 1850.

———. "Recherches électro-physiologiques, pathologiques et thérapeutiques sur le diaphragme." *L'Union médicale*, 1853, *101*, p. 26.

———. "Recherches électro-physiologiques sur les muscles de la face et sur les interosseux." Archives de l'Académie des sciences, Paris, 12 March 1850, ff. 1–24.

———. "Troisième et dernière note sur les fonctions des muscles de la face étudiées à l'aide de la galvanisation localisée." Archives de l'Académie des sciences, Paris, 12 March 1850, f. 15.

Dumas, G. "L'Expression des émotions." In *Traité de psychologie*. Paris: Félix Alcan, 1923, vol. 2, pp. 606–690.

Duval, M. "Discours de M. le Professeur Mathias Duval." In *Inauguration du monument élevé à la mémoire de Duchenne, de Boulogne*. Paris, 1897, pp. 25–30.

Duvivier, M. "Liste des peintres et des sculpteurs couronnés jusqu'en 1861 dans le concours de la tête d'expression." *Archives de l'art français*, Paris, 1861, *1*, pp. 195–208.

Egger, V. "La physiologie cérébrale et la psychologie." *Revue des deux mondes*, 1877, *24*, pp. 193–211.

Félibien, A. "Préface aux Conférences de l'Académie royale de peinture et de sculpture pendant l'année 1667." In *Les Conférences de l'Académie royale de peinture et de sculpture au XVIIe siècle*, ed. A. Mérot. Paris: École nationale supérieur des beaux-arts, 1996, pp. 43–59.

Fontaine, A., *Les doctrines d'art en France*. Paris, 1909.

Foucault, M. *The Birth of the Clinic*. Translated by A. M. Sheridan Smith. New York: Vintage Books, 1973.

———. *Les mots et les choses*. Paris: Gallimard, 1966.

———. *Naissance de la clinique*. Paris: PUF, 1963.

———. *The Order of Things: An Archaeology of Human Sciences*. New York: Vintage Books, 1994.

———. "La peinture photogénique." *Dits et écrits*. Paris: Gallimard, 1994, vol. 2, pp. 707–715.

———. "Photogenic Painting." Translated by P. A. Walker. *Critical Texts*, 1989, *6*, no. 3, pp. 1–2.

Frizot, M., and F. Ducros. *Du bon usage de la photographie: Une anthologie de textes choisis*, 27. Paris: Centre National de la Photographie, 1987.

Goethe, J. W. "L'Essai sur la peinture de Diderot." In *Écrits sur l'art*. Paris: GF-Flammarion, 1996 (1799), pp. 188–240.

———. "On the *Laocoön* Group " (1798). In *Essays on Art and Literature*, ed. John Gearey, trans. Ellen von Nardroff and Ernest H. von Nardroff. New York: Suhrkamp, 1986.

———. "Sur Laocoön." In *Écrits sur l'art* (1798 [JM1]), pp. 164–178.

Gombrich, E. H. *Art and Illusion: A Study in the Psychology of Pictorial Representation.* New York: Pantheon, 1960.

———. *L'Art et l'illusion.* Paris: Gallimard, 1996 (New York and London: Phaidon Press, 1960).

Gratiolet, P. Conférence de M. Gratiolet (soirées scientifiques de la Sorbonne, ed. E. Algrave). "De l'homme et de sa place dans la création." *Revue des cours scientifiques,* March 19, 1864, *16,* pp. 189a–193b.

———. *De la physionomie et des mouvements d'expression.* Paris: J. Hetzel, 1865.

Guilly, P. *Duchenne de Boulogne.* Paris: J.-B. Baillière et fils, 1936.

Guyot, Y. "Psychologie: L'art et la science." *Revue scientifique,* 1887, series 3, *14,* pp. 138–146.

Hacks, Ch. *Le geste.* Paris: Marpon et Flammarion, 1892.

Hans-Drouin, A-M. *La communication non verbale avant la lettre.* Paris: L'Harmattan, 1995.

Heilbrun, F. "L'Art du portrait photographique chez Félix Nadar." In *Nadar: Les années créatrices: 1854–1860.* Paris: Réunion des musées nationaux, 1994, pp. 42–103.

Hogarth, W. *Analyse de la beauté.* Paris: École nationale supérieur des beaux-arts, 1991 (1753).

Hueston, J. T., and R. A. Cuthbertson. "Duchenne de Boulogne and Facial Expression." *Annals of Plastic Surgery* 1978, *1,* p. 441.

Jammes, A. "Duchenne de Boulogne: La grimace provoquée et Nadar." *Gazette des Beaux-Arts,* 1978, 1319, *92,* pp. 215–220.

Jammes, A., and R. Sobieszek. *French Primitive Photography.* Philadelphia: Philadelphia Museum of Art, 1969.

Johnson, D. "Desire Demythologised: David's *L'Amour quittant Psyché.*" *Art History,* 1986, *4,* pp. 450–470.

von Kleist, H. "Sur le théâtre de marionnettes." *Europe,* 1986, no. 686–687, pp. 17–22.

Lacan, E. *Esquisses photographiques à propos de l'exposition universelle.* Paris: Grassart, 1856.

Latour, A. "Mécanisme de la physiologie humaine." *L'Union médicale,* 1862, *103,* pp. 369–375, 417–423.

Lavater, G. "De l'art de voir et d'observer les physiognomies." In *L'Art de connaître les hommes par la physionomie.* 2nd ed. Paris: L.-T. Cellot, 1820, vol. 5, pp. 1–159.

———. "Introduction et considérations générales." In *L'Art de connaître les hommes par la physionomie,* vol. 1, pp. 147–413.

———. "Quelques vues générales sur la physiognomonie." In *L'Art de connaître les hommes par la physionomie,* vol. 3, pp. 169–258.

———. "Rapports de la physiognomonie avec la peinture." In *L'Art de connaître les hommes par la physionomie,* vol. 8, pp. 1–145.

Le Brun, Ch. "Conférence sur l'expression générale et particulière." In *L'Art de connaître les hommes par la physiognomonie.* Paris: L.-T. Cellot, 1820, vol. 9, pp. 261–302.

———. "Lecture on Expression: Conférence sur l'expression générale et particuliére." In *The Expression of the Passions,* ed. and trans. J. Montagu. New Haven and London: Yale University Press, 1994.

———. "Sur le Saint Michel terrassant le démon de Raphaël" (7 May 1667). In *Les Conférences de l'Académie royale de peinture et de sculpture au XVIIe siècle,* ed. A. Mérot. Paris: École nationale supérieur des beaux-arts, 1996, pp. 60–67.

———. "Sur les Israélites recueillant la manne dans le désert de Poussin" (5 November 1667). In *Les Conférences de l'Académie royale,* pp. 98–112.

Lee, R. W. *Ut Pictura Poesis.* Paris: Macula, 1998.

———. *Ut Pictura Poesis: The Humanistic Theory of Painting.* New York: Norton, 1967.

Le Hay (É.-S. Chéron). *Livre à dessiner, composé de têtes tirées des plus beaux ouvrages de Raphaël.* Paris, 1706.

Lemoine, A. *L'Âme et le corps.* Paris, 1862.

———. *La physionomie et la parole.* Paris: Gemer Baillière, 1865.

Lessing, G. E. *Laocoön.* Paris: Hermann, 1990 (1766).

———. *Laocoön: An Essay on the Limits of Painting and Poetry.* Translated by E. A. McCormick. Baltimore: Johns Hopkins University Press, 1984.

Levinas, E. "Ethics of the Face." In *Totality and Infinity: An Essay on Exteriority,* trans. A. Lingis. Pittsburgh: Duquesne University Press, 1969.

McCauley, E. A. *Industrial Madness: Commercial Photography in Paris 1848–1871.* New Haven and London: Yale University Press, 1994.

Montagu, J. *The Expression of the Passions.* New Haven and London: Yale University Press, 1994.

Moreau, J.-L. "Analyse anatomique et physiologique du visage." In *L'Art de connaître les hommes par la physionomie.* 2nd ed. Paris, 1820, vol. 4, pp. 125–313.

———. "Observations physiologiques sur l'expression et les caractères des passions." In *L'Art de connaître les hommes par la physionomie,* vol. 5, pp. 210–269.

Nadar, F. *Quand j'étais photographe.* Arles: Acte Sud, 1998 (1900).

———. *Revendication de la propriété exclusive du pseudonyme Nadar.* Paris, 1857.

Nietzsche, F. "La généalogie de la morale." In *Oeuvres philosophiques complètes.* Paris: Gallimard, 1971, vol. 7, p. 270.

———. "Humain, trop humain." In *Oeuvres philosophiques complètes.* Paris: Gallimard, 1988, vol. 3, p. 299.

————. *Human, All Too Human: A Book for Free Spirits.* Translated by M. Faber. Lincoln: University of Nebraska Press, 1984.

————. *On the Genealogy of Morality.* Translated by C. Diethe. New York: Cambridge University Press, 1994.

————. *The Twilight of the Gods.* New York: Oxford University Press, 1998.

Panofsky, D. "Gilles ou Pierrot? Iconographic notes on Watteau." *Gazette des Beaux-Arts,* 1952, *39,* pp. 319–340.

Parsons, J. "Human Physiognomy Explain'd: In the Crounian Lectures on Muscular Motion. For the Year 1746. Read before the Royal Society." Supplement to the *Philosophical Transactions,* 1747, *44,* pp. 1–77.

de Piles, Roger. *Cours de peinture par principes.* Nîmes: J. Chambon, 1990 (1709).

Pommier, E. *Théories du portrait: De la Renaissance aux Lumières.* Paris: Gallimard, 1998.

Posner, D. "Charles Lebrun's Triumphs of Alexander." *Art Bulletin,* 1959, *91,* pp. 237–247.

Ralph, B. *The School of Raphaël: Or the Student's Guide to Expression in Historical Painting.* London, 1759.

Rayer, P. "Rapport sur l'ouvrage de M. le Docteur Duchenne de Boulogne. Procès verbaux des séances de la commission." Archives Nationales, F.17 3100.1.

Ribot, Th. *La psychologie des sentiments.* 16th ed. Paris: Presses Universitaires de France, 1939.

Richer, P. *Artistic Anatomy.* Edited and translated by R. B. Hale. New York: Watson-Guptill, 1971.

————. *Physiologie artistique.* Paris, 1895, p. 12.

Rouillé, A. *La photographie en France, textes et controverses: Une anthologie 1816–1871.* Paris: Macula, 1989.

Ruskin, J. *Conférences sur l'architecture et la peinture.* Paris: H. Laurens, 1929.

————. *Lectures on Architecture and Painting, Delivered at Edinburgh in November 1853.* New York: Charles E. Merrill, 1892.

Sarlandière, J.-B. *Physiologie de l'action musculaire appliquée aux arts d'imitation.* Paris: De Lachevardière, 1830.

Serres, E. R. A. *Anatomie comparée transcendante.* Paris, 1859.

————. "Rapport de M. Serres sur l'ouvrage de M. Duchenne de Boulogne, intitulé *Mécanisme de la physionomie humaine.*" Archives Nationales, Meeting 26 May 1864, F/17/3101, ff. 1–15.

Sobieszck, R. *Ghost in the Shell.* Cambridge and London: MIT Press, 2000.

Sue, J. J. *Élémens d'anatomie, à l'usage des peintres, des sculpteurs, et des amateurs.* 2nd ed. Paris, 1788.

Testelin, H. "Sur l'expression générale et particulière" (6 June 1675). In *Les Con-*

férences de l'Académie royale de peinture et de sculpture au XVIIe siècle, ed. A. Mérot. Paris: École nationale supérieur des beaux-arts, 1996, pp. 313–326.

Töpffer, R. "Essai de physiognomonie" (1845). In *L'Invention de la bande dessinée*. Textes présentés par T. Groensteen et B. Peeters. Paris: Hermann, 1994, pp. 185–225.

Verneuil, A. "Mécanisme de la physionomie humaine, ou analyse électro-physiologique de l'expression des passions applicable à la pratique des arts plastiques." *Gazette hebdomadaire de médecine et de chirurgie*, 1862, *28*, pp. 445a–448b.

Virey, J.-J. "Physiognomonie." *Dictionnaire des sciences médicales.* Paris: Panckoucke, 1820, vol. 42, pp. 188–228.

Watelet, C.-H., and M. Lévesque. "Détail." *Dictionnaire des arts de peinture, sculpture et gravure.* Paris, 1792, vol. 1, pp. 617–623.

Winckelmann, J. J. *Histoire de l'art chez les anciens.* 2 vols. Amsterdam, E. van Harrevelt, 1766.

———. *History of the Art of Antiquity.* Translated by H. F. Mallgrave. Los Angeles: Getty Research Institute, 2006.

———. *Reflections on the Imitation of Greek Works in Painting and Sculpture.* Translated by E. Heyer and R. C. Norton. La Salle, Ill.: Open Court, 1987.

———. *Réflexions sur l'imitation des oeuvres grecques en peinture et en sculpture.* Nîmes: J. Chambon, 1991 (1755).

Winslow, J.-B. *Exposition anatomique de la structure du corps humain.* Paris: G. Desprez and J. Desseartz, 1732.

Wittgenstein, L. *Philosophical Investigations.* Translated by G. E. M. Anscombe. New York: Macmillan, 1953.

Index

Cultural Memory | *in the Present*

Martin Seel, *Aesthetics of Appearing*

Nanette Salomon, *Shifting Priorities: Gender and Genre in Seventeenth-Century Dutch Painting*

Jacob Taubes, *The Political Theology of Paul*

Jean-Luc Marion, *The Crossing of the Visible*

Eric Michaud, *The Cult of Art in Nazi Germany*

Anne Freadman, *The Machinery of Talk: Charles Peirce and the Sign Hypothesis*

Stanley Cavell, *Emerson's Transcendental Etudes*

Stuart McLean, *The Event and Its Terrors: Ireland, Famine, Modernity*

Beate Rössler, ed., *Privacies: Philosophical Evaluations*

Bernard Faure, *Double Exposure: Cutting Across Buddhist and Western Discourses*

Alessia Ricciardi, *The Ends of Mourning: Psychoanalysis, Literature, Film*

Alain Badiou, *Saint Paul: The Foundation of Universalism*

Gil Anidjar, *The Jew, the Arab: A History of the Enemy*

Jonathan Culler and Kevin Lamb, eds., *Just Being Difficult? Academic Writing in the Public Arena*

Jean-Luc Nancy, *A Finite Thinking*, edited by Simon Sparks

Theodor W. Adorno, *Can One Live after Auschwitz? A Philosophical Reader*, edited by Rolf Tiedemann

Patricia Pisters, *The Matrix of Visual Culture: Working with Deleuze in Film Theory*

Andreas Huyssen, *Present Pasts: Urban Palimpsests and the Politics of Memory*

Talal Asad, *Formations of the Secular: Christianity, Islam, Modernity*

Dorothea von Mücke, *The Rise of the Fantastic Tale*

Marc Redfield, *The Politics of Aesthetics: Nationalism, Gender, Romanticism*

Emmanuel Levinas, *On Escape*

Dan Zahavi, *Husserl's Phenomenology*

Rodolphe Gasché, *The Idea of Form: Rethinking Kant's Aesthetics*

Michael Naas, *Taking on the Tradition: Jacques Derrida and the Legacies of Deconstruction*

Herlinde Pauer-Studer, ed., *Constructions of Practical Reason: Interviews on Moral and Political Philosophy*

Jean-Luc Marion, *Being Given That: Toward a Phenomenology of Givenness*

Theodor W. Adorno and Max Horkheimer, *Dialectic of Enlightenment*

Ian Balfour, *The Rhetoric of Romantic Prophecy*

Martin Stokhof, *World and Life as One: Ethics and Ontology in Wittgenstein's Early Thought*

Gianni Vattimo, *Nietzsche: An Introduction*

Jacques Derrida, *Negotiations: Interventions and Interviews, 1971-1998*, ed. Elizabeth Rottenberg

Brett Levinson, *The Ends of Literature: The Latin American "Boom" in the Neoliberal Marketplace*

Timothy J. Reiss, *Against Autonomy: Cultural Instruments, Mutualities, and the Fictive Imagination*

Hent de Vries and Samuel Weber, eds., *Religion and Media*

Niklas Luhmann, *Theories of Distinction: Re-Describing the Descriptions of Modernity*, ed. and introd. William Rasch

Johannes Fabian, *Anthropology with an Attitude: Critical Essays*

Michel Henry, *I Am the Truth: Toward a Philosophy of Christianity*

Gil Anidjar, *"Our Place in Al-Andalus": Kabbalah, Philosophy, Literature in Arab-Jewish Letters*

Hélène Cixous and Jacques Derrida, *Veils*

F. R. Ankersmit, *Historical Representation*

F. R. Ankersmit, *Political Representation*

Elissa Marder, *Dead Time: Temporal Disorders in the Wake of Modernity (Baudelaire and Flaubert)*

Reinhart Koselleck, *The Practice of Conceptual History: Timing History, Spacing Concepts*

Niklas Luhmann, *The Reality of the Mass Media*

Hubert Damisch, *A Theory of /Cloud/: Toward a History of Painting*

Jean-Luc Nancy, *The Speculative Remark: (One of Hegel's bon mots)*

Jean-François Lyotard, *Soundproof Room: Malraux's Anti-Aesthetics*

Jan Patočka, *Plato and Europe*

Hubert Damisch, *Skyline: The Narcissistic City*

Isabel Hoving, *In Praise of New Travelers: Reading Caribbean Migrant Women Writers*

Richard Rand, ed., *Futures: Of Jacques Derrida*

William Rasch, *Niklas Luhmann's Modernity: The Paradoxes of Differentiation*

Jacques Derrida and Anne Dufourmantelle, *Of Hospitality*

Jean-François Lyotard, *The Confession of Augustine*

Kaja Silverman, *World Spectators*

Samuel Weber, *Institution and Interpretation: Expanded Edition*

Jeffrey S. Librett, *The Rhetoric of Cultural Dialogue: Jews and Germans in the Epoch of Emancipation*

Ulrich Baer, *Remnants of Song: Trauma and the Experience of Modernity in Charles Baudelaire and Paul Celan*

Samuel C. Wheeler III, *Deconstruction as Analytic Philosophy*

David S. Ferris, *Silent Urns: Romanticism, Hellenism, Modernity*

Rodolphe Gasché, *Of Minimal Things: Studies on the Notion of Relation*

Sarah Winter, *Freud and the Institution of Psychoanalytic Knowledge*

Samuel Weber, *The Legend of Freud: Expanded Edition*

Aris Fioretos, ed., *The Solid Letter: Readings of Friedrich Hölderlin*

J. Hillis Miller / Manuel Asensi, *Black Holes / J. Hillis Miller; or, Boustrophedonic Reading*

Miryam Sas, *Fault Lines: Cultural Memory and Japanese Surrealism*

Peter Schwenger, *Fantasm and Fiction: On Textual Envisioning*

Didier Maleuvre, *Museum Memories: History, Technology, Art*

Jacques Derrida, *Monolingualism of the Other; or, The Prosthesis of Origin*

Andrew Baruch Wachtel, *Making a Nation, Breaking a Nation: Literature and Cultural Politics in Yugoslavia*

Niklas Luhmann, *Love as Passion: The Codification of Intimacy*

Mieke Bal, ed., *The Practice of Cultural Analysis: Exposing Interdisciplinary Interpretation*

Jacques Derrida and Gianni Vattimo, eds., *Religion*

Printed in the USA
CPSIA information can be obtained
at www.ICGtesting.com
JSHW021438221024
72172JS00005B/49

9 780804 758512